CONCEPTS OF POLYMER THERMODYNAMICS

POLYMER THERMODYNAMICS LIBRARY, VOL. 2

Concepts of Polymer Thermodynamics

MENNO A. VAN DIJK

Shell Research and Technology Centre, The Netherlands

ANDRÉ WAKKER

Shell Research and Technology Center, Belgium

CRC Press
Taylor & Francis Group
Boca Raton London New York

CRC Press is an imprint of the
Taylor & Francis Group, an **informa** business

CRC Press
Taylor & Francis Group
6000 Broken Sound Parkway NW, Suite 300
Boca Raton, FL 33487-2742

© 1997 by Taylor & Francis Group, LLC
CRC Press is an imprint of Taylor & Francis Group, an Informa business

First issued in paperback 2019

No claim to original U.S. Government works

ISBN-13: 978-0-367-44792-2 (pbk)
ISBN-13: 978-1-56676-623-4 (hbk)

Visit the Taylor & Francis Web site at
http://www.taylorandfrancis.com

and the CRC Press Web site at
http://www.crcpress.com

Table of Contents

Preface

When we started our research jobs on polymer thermodynamics at the Shell research laboratory in Amsterdam we were both new in the areas of polymers as well as thermodynamics, having just finished theses on respectively undercooled water and microemulsions. This allowed us to have a fresh look at the subject. Another circumstance that gave us a non-traditional approach was that our research group was housed in an environment of thermodynamic research. Our neighbors were not synthesizing, blending, extruding or injection molding polymers; they were doing calculations on gas-liquid equilibria in distillation columns, measuring vapor pressures and developing equations of state for hydrocarbon mixtures.

We had to find our way in polymer thermodynamics through many books and articles. Many of these literature sources were either written from a very practical point of view, taking existing theories for granted or from a very theoretical point of view, ignoring the practicalities. It took us much time and long philosophical discussions to understand the physical meaning of the concepts that were used and their relation to "small molecules thermodynamics". We look back with great pleasure to these discussions, often continued after work hours and with a good glass of "Trappist". During these discussions we gradually obtained an understanding of the various fields and 'schools' in this area and their interrelationships. It was a fortunate circumstance that we had a first-class laser light scattering set-up available at the laboratory. This greatly helped us to verify some of our concepts directly on well defined model systems.

Writing this book was a great opportunity to express what we had learnt. In fact, we have set ourselves the goal to write the book that we would have liked to have had before. No doubt we would have found another good reason for the "Trappist". We sincerely hope that this book will prove to be of value to many new or experienced workers in this area of research. If not for its content then for the discussions it may provoke. When some pages may seem overloaded with equations that is because we wished to show intermediate steps in a derivation rather then to confuse or intimidate the reader. In several cases these intermediate steps are not trivial and imply additional assumptions.

As we found out, writing a book is a big project which requires, in addition to a permission to publish, good faith, critical pre-readers, general support, inspiration and patience. For this we acknowledge with pleasure and true gratitude the following:

Permission to publish all this:	Shell Research
Good faith:	our publisher
Critical reading of (parts of) the manuscript:	Peter Hilbers and Alan Batt
Very critical reading of part of the manuscript:	Eric Hendriks
General support:	many friends, relatives and colleagues who kept reminding us of our duties asking 'how the book was going'.
Inspiration:	All the foregoing and in particular all members of 'Ceetje eXtra': Christian Houghton-Larssen, Ferry van Dijk, Valentijn Hommels, and Anthony Lucassen.
Patience:	our wives.

Menno van Dijk
André Wakker

Chapter 1

INTRODUCTION

A large and growing fraction of synthetic materials consists of mixtures of polymers[1,2] and thermodynamics is an indispensable tool in the development of these materials.

These materials are one, two, or multi phase systems. The basic reason for this growing importance is the fact that the commercial introduction of a chemically new polymer that is not very expensive or only useful in a special application has become a rare event. It looks as though we shall have to live with a basic set of polymers as the standard building blocks of our future materials. Therefore, new materials must be obtained by a suitable combination of these building blocks. Primarily, "suitable" means: with an appropriate morphology. Thermodynamics play a key role in understanding why two polymers are miscible or, if not, what morphology they form.

Thermodynamics is a broad subject. It has been developed for the study of heat engines at the end of the nineteenth century. Its framework: the fundamental laws, the definition of temperature and the notion of entropy and free energy proved to be extremely useful to many other branches of physics and chemistry, including the description of the phase behavior of mixtures. Thermodynamics provides the most fundamental rules that must be obeyed by all systems, irrespective of their detailed atomic structure. This is the strength and also the weakness of thermodynamics: its rules are fundamental but it does not provide a method to allow for the effects of chemical detail.

This is where statistical mechanics enters the stage. Statistical mechanics provides us with a means to link the microscopic behavior at the level of atoms and molecules to the macroscopic world where the laws of thermodynamics reign. Usually the whole field is included in the term "thermodynamics".

Only a few polymer pairs are miscible in the thermodynamic sense as relevant to this book and very few have been commercialized. Polystyrene/polyphenylene-ether is the classical example of a successful miscible blend. As we shall see, it is well understandable on polymer thermodynamic grounds that only few polymer pairs are miscible. The entropic driving force towards miscibility (at a molecular scale) is very much smaller in mixtures of polymers than in mixtures

1

of small molecules. Therefore, some special conditions have to be met for a polymer pair to be miscible. A temptation to conclude that the relevance of polymer thermodynamics is that it perfectly explains why one should not waste time studying it, must however be rejected.

Usually, when two chemically different polymers are blended in some mixing equipment, the resulting material will have a coarse morphology with poor interaction between the two phases and consequently inferior mechanical properties. Here polymer thermodynamics may play a role. It can be used to rank polymer pairs in order of immiscibility, acknowledging that miscible pairs are rare. Materials with a higher "thermodynamic immiscibility" have a coarser morphology and more inferior mechanical properties. These coarse blends are often "compatibilized" by adding a suitable third component that acts much like a surfactant. It resides at the interface where it lowers the interfacial tension which leads to a finer structure. Also, the adhesion between both phases is enhanced. These compatibilizers are often (block) copolymers. Thermodynamics may be used to search for suitable chemical structures.

There is more. Morphology plays a key role in determining properties. We have seen this in civil engineering construction where the use of clever shapes (e.g., the H-beam) has enabled the lightweight constructions of bridges and buildings of today compared to the heavy solid constructions of centuries ago. Similarly, intelligent use of appropriate morphologies will play a key role in material engineering. This means that efforts will be needed to develop methods that create the desired morphology.

For example, a technique of growing importance is reactive processing. Here one or more of the constituents may be of low molecular weight and liquid. Other polymeric constituents may be dissolved in this liquid. During the subsequent conversion of the low molecular weight species to polymers, the other components will become immiscible and will form a separate phase. These phase separation processes may be very complicated, even for systems with two components. Thermodynamics is then an essential tool to help understand the observed behavior and to design new materials. High impact polystyrene (HIPS) is an example of a blend that is made by reactive processing (polymerizing a solution of polybutadiene in styrene monomer).

Originally, the interest in the phase behavior of polymeric systems was not geared to the development of new materials but rather to a better understanding of polymers as such. Attention was focused on the behavior of polymer solutions. Polymers have a negligible vapor pressure. Therefore the only way to study more or less isolated polymer molecules is to dissolve them in a liquid. As an additional advantage, polymer solutions are easier to handle than pure polymers with their high viscosities and melting points.

The independent publications of the, later to become famous, expression for the combinatorial entropy of mixing of a polymer solution by Flory and Huggins in 1942 may be seen as a historical starting point. Since then a large number of theories have been developed by various workers with a growing interest in mixtures of polymers. At the same time, a large body of experimental data has been building up in the form of phase diagrams, interaction parameters, scattering functions, etc.

Recently a new discipline has emerged. The advent of powerful computers has led to a whole new branch of "experiments". Molecular modelling, molecular dynamics and Monte Carlo simulations are now increasingly being applied to enhance our understanding of polymer thermodynamic behavior.

In this vast amount of theories, experiments and computer simulations one easily gets "mixed up". It is the purpose of this book to help the reader "demix" and get a feel for the scientific developments in this area. Particular attention has been paid to clarify the relations between the thermodynamics of polymeric mixtures and those of "normal" liquids. Chapter 2 is fully devoted to he thermodynamic description of the phase behavior of low molecular weight molecules. This once served as the framework on which the theories for polymers were build. Some of the concepts introduced in Chapter 2 are almost exclusively applied in polymer science (e.g., solubility parameters). By discussing these topics without reference to polymers we want to emphasize that these notions are not particular to polymer science but of a more general nature.

In Chapter 3 polymers will be introduced and we will discuss the natural extension of the concepts of Chapter 2 to high molecular weight molecules. The notions of Chapter 3, in particular the celebrated Flory-Huggins expression for the free energy of mixing of a polymeric mixture, form the foundations of practically all detailed theories that have been developed. Therefore, they will be rather extensively discussed with particular emphasis on their precise physical meaning.

Many theories along various different lines have been developed to improve the predictive and descriptive capabilities of the Flory-Huggins model. These will be discussed in Chapter 4, again with emphasis on physical meaning rather then mathematical detail and comprehensiveness.

The computer is increasingly used to enhance our understanding of material behavior. Polymer science is no exception. In Chapter 5, the basic relevant computer simulation techniques are introduced. It will be shown that simulation of polymeric systems with chemical detail is not yet feasible. Computer simulations do, however, play a significant role in a qualitative understanding of the often subtle boundaries between miscibility, immiscibility and partial miscibility of polymeric mixtures.

Last but not least, in Chapter 6 ample attention will be given to experimental work. Emphasis will be on those techniques that directly and quantitatively probe the free energy of mixing, particularly scattering techniques. The relevant results from experiments on polymer solutions and blends will be presented, discussed and compared to various thermodynamic model descriptions.

Finally, it must be noted that it was not the ambition of the authors to write a comprehensive overview of all theoretical and experimental know-how. It is the intention of this book to elucidate the physical meaning and practical usefulness of many important developments and their relevance to attempt to predict polymeric phase behavior.

REFERENCES

1.   L. A. Utracki, *Polym. Eng. Sci.*, **35**, 2 (1995).
2.   L. A. Utracki, *Polymer Alloys and Blends, Thermodynamics and Rheology*, Hanser Publishers, Munchen 1989.

Chapter 2

ELEMENTS OF
THERMODYNAMICS OF MIXTURES

2.1 FUNDAMENTALS

2.1.1 THE LAWS OF THERMODYNAMICS

This is not a textbook on thermodynamics,[1,2,3] we thus give only overview of fundamental principles. We will extensively use the tools that thermodynamic and statistical mechanic theories provide. This chapter serves as introduction to the basic framework of thermodynamics and some important relations that will be used later.

The zeroth law is concerned with thermal equilibrium. It states that if two systems are separately in thermal equilibrium with a third, they must be in equilibrium with each other. As an example of the practical consequence, consider a 2-phase polymer solution, consisting of a solvent-rich and a polymer-rich phases in thermal equilibrium with the solvent vapor phase. The zeroth law implies that the vapor pressure will not change after complete removal of one of the coexisting liquid phases. A more fundamental result of the concept of thermal equilibrium as defined by the zeroth law is the definition of temperature (denoted T). The first law is the principle of conservation of energy, where heat is also recognized as a form of energy (U).

The second law can be formulated in many ways. In terms of practical consequences it always creates difficulties: inefficient engines, high electricity bills for the refrigerator and a desk that is always a mess. A more friendly formulation of the second law, due to Kelvin, is that no process is possible whose sole result is the complete conversion of heat into work. The transformation of another quantity is also involved. This quantity was named entropy S (Greek ητροπη = turn, transformation) by Clausius.

2.1.2 THERMODYNAMIC FUNCTIONS AND RELATIONS

From the first and the second law one derives a relation between changes of the internal energy U of a system and changes of its entropy and volume:

$$dU = TdS - PdV \qquad [2.1]$$

where T is the temperature and P is the pressure in the system. The term PdV enters the equation as the work done by the system; that is work done against hydrostatic pressure, P. This is the usual form of the equation. One should however keep in mind that other contributions to the work done by the system exist. Examples are γdA for work done by increasing the surface area by dA against the surface tension γ and FdL for strain against a tension F. With this general character of P and V in mind, we conclude from Eq. (2.1) that the internal energy of a system (with fixed composition) is a function of entropy and volume only: U = U(S,V). If this functional relation is known explicitly, we have complete (thermodynamic) information on the system. For example, the pressure is given by:

$$P = -\left(\frac{\partial U}{\partial V}\right)_S \qquad\qquad [2.2]$$

It can be shown that the thermodynamic equilibrium state for a given S and V corresponds to a minimum of the internal energy U. The equilibrium state of a system with given values for U and V corresponds to a maximum of entropy. By rearranging S, P, V, and T, several other fundamental equations with different independent variables can be derived. For our purposes, entropy and volume as independent variables are not very convenient. In liquid-liquid equilibria one typically uses temperature and pressure as independent variables. In this case, the so-called Gibbs free energy G is the appropriate thermodynamic potential:

$$G = G(P,T) = U + PV - TS = H - TS \qquad\qquad [2.3]$$

where H is called enthalpy. With Eq. (2.1) one derives for the total differential of G:

$$dG = -SdT + VdP \qquad\qquad [2.4]$$

The thermodynamic equilibrium state at specified temperature and pressure corresponds to the minimum of G.

From Eq. (2.4), one obtains:

$$V = \left(\frac{\partial G}{\partial P}\right)_T \qquad S = -\left(\frac{\partial G}{\partial T}\right)_P \qquad\qquad [2.5]$$

If T and V are independent variables, one should consider the Helmholtz free energy F = F(T,V) with

$$F = U - TS \qquad\qquad [2.6]$$

and

$$dF = -SdT - PdV \qquad\qquad [2.7]$$

From Eq. (2.7) and also from Eq. (2.4), one observes that the temperature, pressure and volume of a system cannot be varied independently ($P = -\partial F/\partial V$). The functional relationship between P,T and V can formally be expressed as:

$$f(P,T,V) = 0 \qquad\qquad [2.8]$$

which is referred to as the equation of state of the system.

2.1.3 MIXTURES

In a mixture of different kinds of molecules, the number of molecules n_i of the individual species must also be specified: $G = G(P,T,n_1,...,n_C)$ and:

$$dG = -SdT + VdP + \sum_{i=1}^{C} \mu_i dn_i \qquad\qquad [2.9]$$

where C is the number of components (species). The so-called chemical potential, or partial molar Gibbs free energy, μ_i of component i may thus be defined as:

$$\mu_i = \left(\frac{\partial G}{\partial n_i} \right)_{P,T,n_{j\neq i}} \qquad\qquad [2.10]$$

The chemical potential of a molecule in a mixture is thus given by the increase of the (Gibbs) free energy of the system when one molecule is added to the system and P,T and the numbers of the other species are kept constant under the condition that the system remains in thermodynamic equilibrium. The chemical potential is a so-called intensive property of the system, that is independent of the size of the system. The free energy $G(P,T,n_1,...,n_C)$ is an extensive property. G is proportional to the total mass in the system: $G(P,T,\lambda n_1,...,\lambda n_C) = \lambda G(P,T,n_1,...,n_C)$. With a mathematical theorem by Euler, one may write for such so-called homogeneous functions:

$$G = \sum_{i=1}^{C} n_i \left(\frac{\partial G}{\partial n_i} \right)_{n_{j\neq i}} = \sum_{i=1}^{C} n_i \mu_i \qquad\qquad [2.11]$$

Differentiation of Eq. (2.11) yields:

$$dG = \sum n_i d\mu_i + \sum \mu_i dn_i \qquad\qquad [2.12]$$

For constant T and P, Eq. (2.9) and Eq. (2.11) yield:

$$\sum n_i d\mu_i = 0 \qquad\qquad [2.13]$$

This is a so-called Gibbs-Duhem relation. It states that the chemical potential of all C components cannot vary independently. Basically this is a result of the fact that the chemical potential μ_i is an intensive quantity which implies that it depends only upon the composition of the system and the composition is fully characterized by a set of C-1 mole fractions (at fixed P,T).

The chemical potential has been introduced as the partial molar Gibbs free energy here but could equally well have been defined on the basis of U, H or F:

$$\mu_i = \left(\frac{\partial G}{\partial n_i}\right)_{P,T} = \left(\frac{\partial U}{\partial n_i}\right)_{S,V} = \left(\frac{\partial H}{\partial n_i}\right)_{S,P} = \left(\frac{\partial F}{\partial n_i}\right)_{T,V} \qquad [2.14]$$

For completeness another frequently used expression of the chemical potential needs to be introduced. The chemical potential is expressed in units of energy. It is often convenient to use a dimensionless function, the absolute activity λ_i, defined by:

$$\mu_i = RT \ln \lambda_i \qquad\qquad [2.15]$$

where R is the molar gas constant ($R \approx 8.314$ J K^{-1} mol^{-1}).

2.1.4 COMPOSITION VARIABLES

A mixture is any phase containing more than one component. Mixtures may be gas, dense fluid, liquid, or solid. Since this book focuses on polymeric systems we will restrict ourselves to the liquid mixtures. The first step is to characterize the mixture by specifying the composition. One way to do this is by means of the mole fraction, denoted x. If the mixture consists of n_i molecules labeled i, with i=1...C, then the mole fraction of x_i is given by:

$$x_i = \frac{n_i}{\displaystyle\sum_{i=1}^{C} n_i} \qquad\qquad [2.16]$$

It is often convenient to specify the mass fraction W_i which is related to the mole fraction by:

$$W_i = \frac{x_i M_i}{\displaystyle\sum_{i=1}^{C} x_i M_i} \qquad\qquad [2.17]$$

where M_i denotes the molecular weight of molecule i.

Another important composition variable is the volume fraction, Φ. This is an experimentally less accessible variable as the total volume of the mixture (at constant pressure) is a function of temperature and may not be constant for all mixing ratios. If one assumes zero volume change on mixing then the following relation between volume fraction and weight fraction can be derived:

$$\Phi_i = \frac{\dfrac{W_i}{\rho_i}}{\sum\limits_{i=1}^{C} \dfrac{W_i}{\rho_i}} \qquad [2.18]$$

where ρ_i is the pure component mass density. The concentration c_i, expressed in mass per unit volume is given by:

$$c_i = \Phi_i \rho_i \qquad [2.19]$$

The change of volume on mixing cannot always be neglected. For example, if 100 cm^3 of water is mixed with 100 cm^3 of ethanol at 25°C, the total volume of the mixture is not 200 cm^3 but about 190 cm^3. These effects can be described by the partial molar volume $\overline{V}_i$:

$$\overline{V}_i = \left(\frac{\partial V}{\partial n_i}\right)_{T,P,n_{i=1}} \qquad [2.20]$$

In general $\overline{V}_i$ depends on the composition. If there is no volume change on mixing, $\overline{V}_i$ is constant and equal to the molar volume V_i of pure molecule i. For a C component system one has for the total volume:

$$V = \sum_{i=1}^{C} n_i \overline{V}_i \qquad [2.21]$$

Partial molar quantities are defined as partial derivatives with respect to the number of particles n at fixed P and T, (see Eqs. (2.10) and (2.20)), while the expressions for the extensive variable are often given in mole fractions x, weight fractions W, or volume fractions Φ. It would therefore be convenient to have expressions for these partial molar quantities in terms of derivatives with respect to composition variables instead of numbers. We shall now give a derivation of the relevant expressions for the chemical potential but the results apply to any partial molar quantity.[3] In the derivation below we will use volume fractions and then later give the result for mole and weight fractions. The volume fraction Φ_k is given by:

$$\Phi_k = \frac{n_k V_k}{V} \qquad [2.22]$$

where the partial molar volumes $\overline{V}_k$ were assumed to be constant and equal to their pure component values V_k. Contrary to the numbers of molecules, the volume fractions are not independent: they add up to 1. We (arbitrarily) take $\Phi_2,...,\Phi_C$ as independent variables and use $\Phi_1 = 1 - \Sigma\Phi_{j>1}$. In terms of these independent variables, the Gibbs free energy per unit volume G^V is given by:

$$G^V \equiv \frac{\sum\limits_{i=1}^{C} n_i \mu_i}{V} = \sum_{i=1}^{C} \frac{\Phi_i \mu_i}{V_i} = \left(1 - \sum_{i=2}^{C} \Phi_1\right)\frac{\mu_i}{V_1} + \sum_{i=2}^{C} \Phi_i \frac{\mu_i}{V_i} \qquad [2.23]$$

Differentiation of Eq. (2.23) with respect to $\Phi_{j\neq1}$ with all other independent volume fractions fixed (indicated by Φ) yields:

$$\left(\frac{\partial G^V}{\partial \Phi_i}\right)_\Phi = \frac{\mu_i}{V_i} - \frac{\mu_1}{V_1} \qquad [2.24]$$

The Eqs. (2.23) and (2.24) can be solved for μ_1 and μ_j:

$$\frac{\mu_1}{V_1} = G^V - \sum_{i=2}^{C} \Phi_i \left(\frac{\partial G^V}{\partial \Phi_i}\right)_\Phi \qquad [2.25]$$

and:

$$\frac{\mu_j}{V_j} = G^V - \sum_{i=2}^{C} \Phi_i \left(\frac{\partial G^V}{\partial \Phi_i}\right)_\Phi + \left(\frac{\partial G^V}{\partial \Phi_j}\right)_\Phi \qquad [2.26]$$

for j=2,...,C. For two components (C=2) the equations read:

$$\frac{\mu_1}{V_1} = G^V - \Phi_2 \frac{\partial G^V}{\partial \Phi_2} \qquad [2.27]$$

and:

$$\frac{\mu_2}{V_2} = G^V + (1 - \Phi_2)\frac{\partial G^V}{\partial \Phi_2} \qquad [2.28]$$

With $\Phi_1 = 1 - \Phi_2$ and $\partial/\partial\Phi_1 = -\partial/\partial\Phi_2$ one may establish symmetry between component 1 and 2 in Eqs. (2.27-2.28), such that the effect of the arbitrary choice of Φ_1, as dependent composition variable, disappears.

Similar relations can be derived for mole and mass fractions as composition variables. For mole fractions, replace the V_i by 1 and G^V (=G/V) by G/N where N is the total number of molecules. For weight fractions, replace the V_i by M_i and G^V by G/W where W is the total mass of the system.

2.2 PHASE EQUILIBRIA

2.2.1 PHASES

Up to this point, the 'system' was seen as some black box characterized by volume, pressure, temperature, energy, entropy, composition, etc. We have seen that the thermodynamic equilibrium state of 'the system', when pressure and temperature are specified, corresponds to a minimum of the Gibbs free energy. In the spirit of this chapter, which deals with general thermodynamic concepts, we have not gone into any detail of how such a state may look like. Now, we must introduce an important experimental observation that needs to be incorporated into the general thermodynamic framework. One observes that a system may consist of two or more states of matter (phases) in thermal and mechanical equilibrium. Many types of such equilibria exist. Examples are: vapor-liquid, liquid-solid, liquid-liquid, liquid crystalline-isotropic and many others. Thermal equilibrium implies equal temperature in both phases, mechanical equilibrium implies equal pressure in both phases. Chemical equilibrium implies equal chemical potentials in both phases as we will now show. We assume that chemical reactions do not take place.

For a transfer of n_i particles of a substance between two phases (I) and (II) at the same temperature T and pressure P, the change in Gibbs free energy is:

$$\partial G = (\mu_i^I - \mu_i^{II})\partial n_i \qquad\qquad [2.29]$$

At thermodynamic equilibrium G is at a minimum, so this variation is zero:

$$\partial G = 0 \qquad\qquad [2.30]$$

With Eq. (2.29), we then obtain for the equilibrium conditions:

$$\mu_i^I = \mu_i^{II}; \quad i = 1,...,C \qquad\qquad [2.31]$$

where C is the number of different components. One may thus calculate the thermodynamic equilibrium state of the two-phase system, either by directly minimizing the free energy function G, or by solving equation (2.31).

Note that we have ignored the role of the physical interface between the phases. The above derivation is valid in the thermodynamic limit of infinite volumes with negligible interfaces. In practice, the interface plays a role in determining the time that is needed to reach the thermodynamic equilibrium state. For example, super saturation of vapor is a result of the fact that it takes energy to create the vapor liquid interface for condensation, and may delay condensation considerably. However, if one waits long enough, there will be some fluctuation to overcome this barrier and condensation will occur. Nevertheless, it is possible to create a system where the interface is so large that it also plays a role in determining the thermodynamic equilibrium state. In such cases, it should be taken into account (see section 2.1.2). After this practical remark we will now return again to basic thermodynamics and discuss some general rules of phase behavior.

2.2.2 GIBBS PHASE RULE

The general equilibrium condition for a system consisting of an arbitrary number of phases and substances is the equality of the chemical potential of each substance in all phases:

$$\mu_i(I) = \mu_i(II) = \dots \mu_i(Ph); \quad I = 1,\dots,C \qquad [2.32]$$

where C is the number of components and Ph is the number of phases in equilibrium. Eq. (2.32) constitutes a set of $(Ph - 1)C$ equations. The number of variables needed to characterize the system is $2 + Ph(C - 1)$ (temperature, pressure and $C - 1$ mole fractions for each phase). The difference between the number of condition variables and the number of chemical potential restraints is the number of degrees of freedom F of the system, i.e., the number of variables that may be independently chosen:

$$F = 2 + Ph(C - 1) = C + 2 - Ph \qquad [2.33]$$

Equation (2.33) expresses the Gibbs phase rule. It states for example that in a one component system (C=1), there can at most be three phases (Ph=3) in equilibrium but only at one certain well defined temperature and pressure (F=0). Such a state is called a triple point. In a two component system there can at most be four phases in equilibrium and at a given pressure, e.g. atmospheric pressure, at most three. A P,T diagram is a graph with a pressure and temperature axis, indicating the thermodynamic equilibrium states at each point P,T. The phase rule tells us that a state of two coexisting phases may exist in a region of the P,T diagram (2 degrees of freedom). Three coexisting phases can only exist on a line (one degree of freedom) and four coexisting phases can only exist at one particular P,T point.

2.2.3 FREE ENERGY OF MIXING

The Gibbs free energy of a mixture of two components is given by:

$$G = n_1\mu_1 + n_2\mu_2 \qquad [2.34]$$

The absolute values of thermodynamic functions such as the entropy and the free energy are irrelevant for the sort of problems discussed in this book. Processes are driven by differences of thermodynamic functions between different states. Thermodynamic equilibrium corresponds to the minimum of free energy, the value of the minimum is immaterial. For mixtures, when the relevant problems are associated with the miscibility of the molecules, it is most convenient to define the pure substances as reference state. We then have for the Gibbs free energy of mixing $\Delta G_M = G_{mixture} - G_{pure}$:

$$\Delta G_M = n_1\Delta\mu_1 + n_2\Delta\mu_2 \qquad [2.35]$$

where $\Delta\mu_i$ is the chemical potential difference of a molecule of component i between the mixture and the reference state:

$$\Delta\mu_i = \mu_i(\text{mixture}) - \mu_i(\text{pure}) \qquad [2.36]$$

In a similar way, the entropy of mixing ΔS_M and the enthalpy of mixing ΔH_M can be defined. The Gibbs free energy of mixing can thus be expressed as:

$$\Delta G_M = \Delta H_M - T\Delta S_M \qquad [2.37]$$

This book deals with the subject of miscibility in polymeric (liquid) systems. The thermodynamic states that will be considered are usually liquid mixtures. The compositions of the mixtures and the conditions under which one or more phases co-exist are the sort of quantities to be calculated. Such problems can be solved completely from a knowledge of the Gibbs free energy of mixing. Therefore, theories focus on developing expressions for $\Delta G_M(P,T,\Phi)$.

For the calculation of liquid-liquid phase behavior, all references to G and μ in the previous discussions may therefore be replaced by ΔG_M and $\Delta\mu$.

The enthalpy of mixing ΔH_M is the heat that is consumed as a result of the mixing of the components at constant pressure. If heat is liberated, the enthalpy of mixing is negative. Exothermic mixing thus gives a negative contribution to the Gibbs free energy of mixing. Negative contributions to ΔG_M drive the system to miscibility as the thermodynamic equilibrium state corresponds to the minimum of ΔG_M. The enthalpy of mixing can be split into two contributions: the change of the internal energy ΔU of the system and the work done by the pressure if there is a volume change ΔV_M of mixing :

$$\Delta H_M = \Delta U_M + P\Delta V_M \qquad\qquad [2.38]$$

The work done by the pressure constitutes the difference between the Gibbs free energy of mixing ΔG_M and de Helmholtz free energy of mixing ΔF_M. If there are no volume changes, the Gibbs and Helmholtz free energies of mixing are identical.

2.2.4 PHASE STABILITY

Consider a binary mixture of n_1 molecules of type 1 and n_2 molecules of type 2, $N = n_1 + n_2$. Let us describe the composition by the volume fraction Φ of, say, component 1. For ease of notation and because the choice of component 1 is arbitrary, the subscript 1 will be omitted. Whether these molecules will form one homogeneous, thermodynamically stable phase at a certain temperature T and pressure P is determined by the free energy difference between the pure phases and the mixture. If this free energy of mixing $\Delta G_M(P,T,\Phi)$ is positive then the molecules will certainly not form a thermodynamically stable mixture because the two pure phases constitute a state with a lower free energy (at this particular P,T and Φ).

If $\Delta G_M(P,T,\Phi) < 0$, this does not necessarily imply that a homogeneous mixture with composition Φ is the thermodynamic equilibrium state. A phase equilibrium of two phases with different compositions may have an even lower free energy (the phase rule excludes more than two phases in equilibrium at arbitrary P and T). As the free energy of mixing is known for all compositions, these compositions can be found by minimization of the total Gibbs free energy of mixing as will now be shown.

Let $G^V(P,T,\Phi)$ be the Gibbs free energy of mixing per unit volume: $G^V \equiv \Delta G_M/V$ (or the total Gibbs free energy per unit volume, the choice is irrelevant for this problem). For a system of two phases with different compositions (denoted Φ^I and Φ^{II}) one has:

$$G^V = v G^V(\Phi^I) + (1 - v) G^V(\Phi^{II}) \qquad\qquad [2.39]$$

where v denotes the phase volume fraction of phase I, i.e., the fraction of the total volume V that is filled with phase I. Again, we assume zero volume change on mixing. Conservation of volume is expressed by:

$$\Phi^F = v\Phi^I + (1 - v)\Phi^{II} \qquad\qquad [2.40]$$

where Φ^F is the fixed overall volume fraction (F from "Feed"). This leads directly to the lever rule which expresses the relative phase volumes in terms of phase volume fractions:

$$v = \frac{\Phi^F - \Phi^{II}}{\Phi^I - \Phi^{II}} \qquad [2.41]$$

By eliminating v from Eq. (2.39) and Eq. (2.40), we derive for the free energy of the two phases:

$$G^V = \frac{G^V(\Phi^I) - G^V(\Phi^{II})}{\Phi^I - \Phi^{II}} \Phi^F + \frac{\Phi^I G^V(\Phi^{II}) - \Phi^{II} G^V(\Phi^I)}{\Phi^I - \Phi^{II}} \qquad [2.42]$$

The meaning of Eq. (2.42) can best be illustrated by a graphical representation of $G^V(\Phi)$ as shown in Figure 2.1. Eq. (2.42) expresses that the free energy of two coexisting phases with compositions of respectively Φ^I and Φ^{II} and overall composition Φ^F can be found by drawing a line between the corresponding points in a graph $G(\Phi)$ (P^I and P^{II} in Figure 2.1) and taking the value of G^V on this line at the overall composition Φ^F (R in Figure 2.1). It follows immediately that the minimum of G^V for a given overall composition Φ^F is obtained in a system with two phases with compositions Φ_S^I and Φ_S^{II} that follow from drawing the common tangent ST in Figure 2.1.

If such a common tangent with $\Phi_S^I < \Phi < \Phi_S^{II}$ cannot be found, systems with overall composition Φ are thermodynamically stable (e.g., point U).

According to Eq. (2.31), the thermodynamic equilibrium compositions can also be found from the condition of equal chemical potentials. With Eqs. (2.27) and (2.28) for the chemical potential in terms of volume fraction derivatives, one then obtains:

$$G^V(\Phi^I) - \Phi^I \left(\frac{\partial G^V}{\partial \Phi}\right)_I = G^V(\Phi^{II}) - \Phi^{II} \left(\frac{\partial G^V}{\partial \Phi}\right)_{II} \qquad [2.43]$$

$$G^V(\Phi^I) + (1 - \Phi^I) \left(\frac{\partial G^V}{\partial \Phi}\right)_I = G^V(\Phi^{II}) + (1 - \Phi^{II}) \left(\frac{\partial G^V}{\partial \Phi}\right)_{II}$$

It can easily be shown by adding and subtracting that these equations are equivalent to the above common tangent construction.

The above analysis implies that a homogeneous mixture with composition $\Phi_S < \Phi < \Phi_T$ is not thermodynamically stable. The analysis yields the thermodynamic equilibrium coexisting phase compositions Φ^I and Φ^{II} but gives no information on how this state will be reached. As discussed before, we may not expect information on the dynamics from equilibrium thermodynamics. Nevertheless, there is

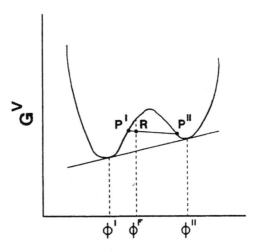

Figure 2.1 Graph of G^V versus composition Φ illustrating how the coexisting phase compositions Φ^I and Φ^{II} can be found graphically.

some more information to be obtained from the expression for ΔG_M regarding the stability against small composition fluctuations. The equilibrium compositions Φ^I and Φ^{II} may be very different and may not be probed by small Brownian motion induced composition fluctuations around Φ. A small composition fluctuation can be described by a local phase separation into two phases with compositions Φ' and Φ'' both close to Φ.

As illustrated in Figure 2.2, the free energy of such a composition fluctuation increases if the curvature of $G^V(\Phi)$ is positive and decreases if the curvature is negative (neglecting interface effects). In the latter case, the composition fluctuation is thermodynamically more favorable and phase separation will proceed. In the first case of a positive curvature, small composition fluctuation have a higher free energy and there will be a thermodynamic driving force back to the original composition Φ. We conclude that the sign of the second derivative of G^V with respect to the composition Φ (the curvature) determines the thermodynamic stability of the system:

$$\frac{\partial^2 G^V}{\partial \Phi^2}(\Phi,P,T) \quad \begin{array}{ll} <0 & \text{unstable} \\ = 0 & \text{spinodal} \\ >0 & \text{(meta)stable} \end{array} \qquad [2.44]$$

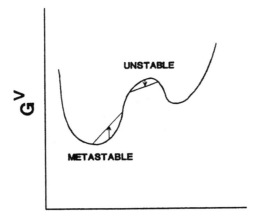

Figure 2.2. Free energy curve showing composition fluctuations in the metastable and unstable regions.

Note that a negative second derivative definitely implies that the mixture is thermodynamically unstable whereas a positive sign may imply a stable or a metastable system.

The special points P,T,Φ where $\partial^2 G^V/\partial\Phi^2$ is zero are called spinodal points. If the system is on a spinodal point there is no thermodynamic driving force to either oppose composition fluctuations or drive them to a macroscopic phase separation. Consequently, large and long range composition fluctuations are possible. This fact makes spinodal points experimentally accessible as these large composition fluctuations cause a large increase in the scattering power of the system. This will be discussed in more detail in Chapter 6.

Here we note that the above derivation was based on the Gibbs free energy per unit volume and compositions expressed in volume fractions. The same reasoning may be applied to the free energy per unit mass and compositions expressed in weight fractions or the free energy per mole of molecules and the composition expressed in mole fractions. The original result, due to Gibbs,[4,5] some 120 years ago, for a two component system is expressed as:

$$\left(\frac{\partial^2 G}{\partial n_1^2}\right)_{P,T,n_2} = 0 \qquad [2.45]$$

In the above derivation, volume fractions were used because the Gibbs free energy of mixing of polymeric mixtures is often expressed in volume fractions. If volume changes on mixing are involved, volume fractions are rather inconvenient quantities and mole fractions are preferred. In terms of mole fractions x, the spinodal criterion reads:

$$\left(\frac{\partial^2 G^N}{\partial x^2}\right)_{P,T} > 0 \qquad\qquad [2.46]$$

If pressure and temperature are specified, G^N is only a function of the composition x. The volume V of the mixture is also fully determined by the composition. However, the free energy may contain contributions which explicitly depend on the volume. Therefore, it is interesting to separate the left hand side of Eq. (2.46) into incompressible (fixed volume) and compressible contributions. By formally writing $G^N(P,T,x) = G^N(P,T,V(x),x)$, the following expression can be derived after some manipulation of derivatives:[6]

$$\frac{d^2 G^N}{dx^2} = \left(\frac{\partial^2 G^N}{\partial x^2}\right)_V - \frac{\left(\frac{\partial^2 G^N}{\partial x \partial V}\right)^2}{\left(\frac{\partial^2 G^N}{\partial V^2}\right)} \qquad\qquad [2.47]$$

Since G is at a minimum, the second derivative of G with respect to V in the above equation is positive. This then leads to the important conclusion that the compressibility term in Eq. (2.47), i.e., the second term, always tends to destabilize the mixture (its contribution to the second derivative is always negative). In fact, the following thermodynamic relation exists for the second derivative:

$$\left(\frac{\partial^2 G}{\partial V^2}\right)_{T,P,x} = \frac{1}{\beta V} \qquad\qquad [2.48]$$

where $\beta \equiv -(\partial \ln V / \partial P)_{T,x}$ is the isothermal compressibility. Usually the compressibility increases with increasing temperature and this means that the destabilizing effect of the compressibility term increases with increasing temperature which may lead to phase separation at higher temperatures (occurrence of a lower critical solution temperature, LCST, see section 2.2.5).

2.2.5 PHASE DIAGRAMS

A phase diagram is a graphic representation of the thermodynamic equilibrium state of a system as a function of a number of relevant parameters (pressure, temperature, volume, composition, electric field, ...). Since graphical representations are limited to three dimensions at the most, one has to keep all these parameters constant except for two or three. Well known phase diagrams are PT and PV diagrams of a single component where regions of vapor, liquid, or vapor-liquid

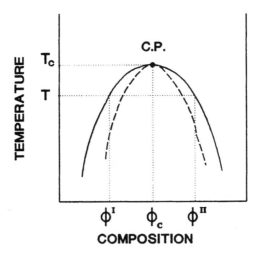

Figure 2.3. Example of T – x phase diagram.

coexistence are indicated. In this book we will mainly be dealing with liquid-liquid equilibria of mixtures with two or more components. The phase behavior of systems that do not involve gas phases is most sensitive to temperature and composition and less sensitive to pressure. Therefore, we shall assume a constant pressure in this section.

The phase behavior of a binary system is then fully described by a temperature - composition diagram. In the previous section, it was shown that such a phase diagram should indicate regions of thermodynamic stability, metastability and instability. An example of such a phase diagram is shown in Figure 2.3. The dashed line is formed by the mathematical solutions Φ,T of the spinodal condition, Eq. (2.46) and is called the spinodal. The full drawn line consists of the solutions Φ^I and Φ^{II} at temperature T of Eq. (2.43). This line is called the binodal. The word binodal reflects the common tangent construction, discussed in Section 2.2.4, while the word spinodal is derived from the mathematical term for a point of inflection (spinode).

From the graphical construction one observes that the spinodal compositions will always be within the binodal. One also sees the following. If the binodal and spinodal compositions come closer (in the above example with increasing temperature), there is inevitably, a temperature where all four compositions are identical. This is the top of the phase diagram in Figure 2.2. This important point is called a critical point (Φ_c, T_c) of the phase diagram. For $T \to T_c$, the second derivatives of ΔG_M with respect to the composition, for $\Phi \uparrow \Phi_c$ and Φ $\downarrow \Phi_c$ are equal (to zero). Therefore, the third derivative also vanishes. The mathematical condition for the liquid-liquid critical point of a binary mixture is thus:

$$\frac{\partial^2 \Delta G_M^V}{\partial \Phi^2} = 0 \qquad\qquad\qquad [2.49]$$

$$\frac{\partial^3 \Delta G_M^V}{\partial \Phi^3} = 0$$

Similar expressions exist for mole fractions or numbers as composition variables. Just like on the spinodal, the system shows infinitely large composition fluctuations at the critical point. Unlike the spinodal points however, the critical point can be approached infinitely close from within the thermodynamically stable region. Furthermore, systems behave similar in many respects close to the critical point, where the second and third free energy derivatives vanish. Their behavior can be described by universal scaling laws. Hence the large theoretical and experimental interest in the critical point.

The existence of critical points depends on the specific properties of the Gibbs free energy of mixing function $\Delta G_M(\Phi, T)$. There is no thermodynamic argument that phase diagrams should have critical points.

Figure 2.4a is a phase diagram which shows the so-called upper critical solution behavior. The temperature T_c is called the upper critical solution temperature (UCST). It is the highest temperature at which two phases may be observed under suitable conditions. The phrase 'upper critical solution temperature' causes confusion because one could rightfully argue that it is also the lowest temperature at which the system is still miscible at all concentrations and that a term like lower critical solution is more appropriate. However, custom has it the other way. Lower critical solution behavior with a lower critical solution temperature (LCST) corresponds to phase diagrams like the one shown in Figure 2.4b. Here phase separation will be observed on raising the temperature. As we will see, LCST phase behavior is quite common in polymeric mixtures, particularly in polymer blends.

Many systems exhibit both LCST and UCST behavior. An example is shown in Figure 2.4c. For temperatures between the UCST and the LCST, the system is miscible in all proportions. At higher and lower temperatures, compositions exist that are not thermodynamically stable. The other extreme, a closed phased diagram as shown in Figure 2.4d is also observed. Another type of phase diagram is shown in Figure 2.4e. Such phase diagrams are often referred to as hour glass phase diagrams. They are the most boring and the most common

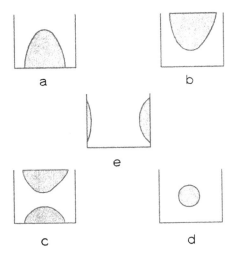

Figure 2.4. Various types of phase diagrams.

phase diagram. They describe systems that are only miscible when one of the components is dilute.

The above discussion covers the basic principles of phase diagrams of binary mixtures. For the treatment of polymeric mixtures in chapter 3, some additional remarks have to be made concerning multicomponent mixtures and the presence of a solid (crystalline) phase. These topics will now be discussed.

2.2.6 MULTICOMPONENT MIXTURES

If more than two components are involved, one quickly runs out of axes for a graphical representation of the phase diagram. One has to plot cross sections or projections of the multidimensional phase diagram at conditions of constancy of some of the independent variables.

Compositions in ternary systems are conveniently represented in a triangular diagram[3] (Figure 2.5). From geometry one learns that the sum of the distances AX+BX+CX = H in Figure 2.5. In a triangle of unit height (H=1) these distances can be used to represent the composition variables x_1, x_2 and x_3 with $x_1 + x_2 + x_3 = 1$ as shown in Figure 2.5.

Figure 2.6 shows a sample phase diagram with a binodal curve of a ternary mixture with the so-called tie lines (or connodals). Along a tie line, all relative proportions of both phases can be found, from zero (i.e., infinitesimally small) at one end to 1 (infinitesimally close to 1) at the other. Contrary to the binary case, the composition of the

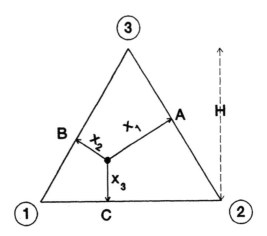

Figure 2.5. Composition triangle.

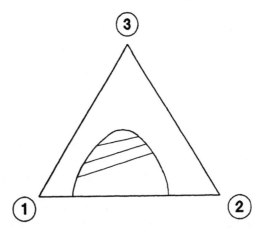

Figure 2.6. Liquid-liquid phase diagram of a ternary mixture at constant pressure and temperature.

incipient phase cannot be read directly from a phase diagram (without tie lines) but should be calculated. Figure 2.6 is the phase diagram at constant pressure and temperature. A third axis may be added to show, for example, the temperature dependence. This is illustrated in Figure 2.7.

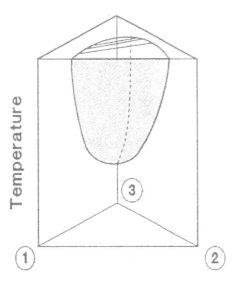

Figure 2.7. Liquid-liquid phase diagram of a ternary mixture at constant pressure.

2.2.7 CRYSTALLIZABLE COMPONENTS

A liquid mixture may not only phase separate by liquid-liquid demixing but also by the crystallization of a component. Crystallization is a first order phase transition and is accompanied by the release of heat, the so-called heat of fusion H_f. On the other hand, crystallization implies an increase of order, that is a decrease of entropy S_f. The melting point T_m is the temperature where both counteracting contributions to the Gibbs free energy $G = H - TS$ are equal (hence ΔG of the phase transition $= 0$):

$$T_m = \frac{H_f}{S_f}$$
[2.50]

If the crystalline solid is brought into contact with a liquid it may or may not dissolve. If it dissolves (partly), there is additional entropy of mixing. So the entropy loss associated with crystallization increases and the melting point is lowered with respect to that of the pure crystalline material. The entropy of mixing ΔS_M is the main contribution to this so-called melting point depression effect. If there is negative (favorable) heat of mixing ΔH_M, the melting point may be lowered even more. If there is a positive heat of mixing the melting point is less depressed then on the basis of the entropy alone. If $\Delta H_M > T_m \Delta S_M$, there is no thermodynamic driving force for the dissolution of the crystalline material and the melting point will not be depressed.

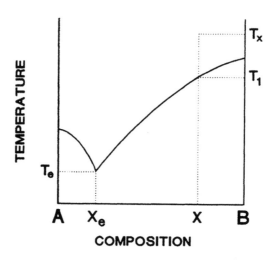

Figure 2.8. Phase diagram of a mixture of two crystalline components.

The general phase diagram of a liquid mixture of two components with two different melting points looks as shown in Figure 2.8.
The lowest point of the phase diagram is called the eutectic point. At
this unique point there are 3 phases in equilibrium: solid A, solid B,
and the liquid mixture of A and B. According to the phase rule, this is
indeed the maximum number of phases for a binary system at arbitrary pressure. Consider the mixture with composition X and temperature T_x as illustrated in Figure 2.8. If the solution is cooled,
precipitation of solid B material will start to occur at $T = T_1$. The
composition of the mixture will follow the melting line (also called
solubility curve) until the eutectic temperature T_e is reached. Further
cooling results in simultaneous precipitation of A and B, such that the
composition of the liquid mixture remains the same, the eutectic
composition X_e. The eutectic composition is analogous to the azeot-
ropic composition in distillation.

Another possibility, which is relevant to polymer solutions is the
simultaneous occurrence of liquid-liquid demixing and melting point
depression phenomena. Such a phase diagram may look like Figure
2.9. If the homogenous liquid solution at composition X_1 and temperature T_1 is cooled, one will observe the occurrence of two liquid phases
when the temperature drops below $T = T_b$. On further cooling the
compositions of the coexisting phases follow the binodal until the
temperature T_{e1} is reached and component B is on the verge of
precipitation. This is again an invariant point with three coexisting

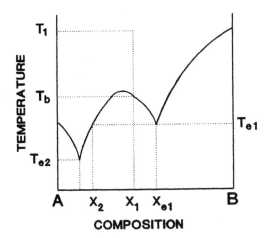

Figure 2.9. Phase diagram of a mixture of two crystalline components that also show liquid-liquid immiscibility.

phases: solid B and 2 liquid solutions with compositions X_2 and X_{e1}. On further cooling, the concentrated B phase of the liquid liquid equilibrium precipitates until the composition of the dilute B phase is reached and we have an equilibrium of solid B and liquid with composition X_d. Finally another invariant point is reached at temperature T_{e2}. For $T < T_{e2}$ one has two solid phases.

2.3 IDEAL MIXTURES

The starting point of the development of thermodynamic theories for gases and vapors was the concept of the ideal gas. In an ideal gas, there are no interactions between the particles. The equation of state of an ideal gas is the well known law of Boyle-Gay Lussac:[1-3]

$$PV = NRT \qquad [2.51]$$

where N is the number of moles and R is the molar gas constant. Since there are no interactions between the gas molecules, the total energy U of the system is independent of the inter particle distances, hence:

$$\left(\frac{\partial U}{\partial V}\right)_T = 0 \qquad [2.52]$$

The entropy S(V,T) of an ideal gas can be derived from its equation of state and is given by:

$$\frac{S}{N} = R \ln V + C_V \ln T + R \ln a \qquad [2.53]$$

for not too low temperature. For extremely low values of the temperature, quantum effects become important, C_V is no longer constant and Eq. (2.53) is no longer valid. Nernst theorem, also referred to as the third law of thermodynamics states that the entropy of any system vanishes at $T = 0$. This essentially means that at $T = 0$, there is only one state of the system, namely the state with the lowest possible energy. With this theorem and using quantum mechanics one may calculate the constant a in Eq. (2.53). For the simplest case of a monoatomic gas ($C_V = 3R/2$) one finds:[7]

$$a = \left(\frac{2\pi mk}{h^2} \right)^{3/2} \qquad\qquad [2.54]$$

where m is the mass of a gas atom, h is Planck's constant and k is Boltzmann's constant ($R = kN_A$). Since there are no interactions between the particles, a mixture of ideal gases obeys the same equation of state, where N now is the total number of molecules. Another way to put this is by means of Daltons law which states that the pressure is the sum of the partial pressures P_i of the different species:

$$P = \sum P_i \qquad\qquad [2.55]$$

and the partial pressures are given by:

$$P_i = \frac{N_i RT}{V} \qquad\qquad [2.56]$$

Consider a vessel divided by a separation into a volume V_1 with N_1 ideal gas molecules and a volume V_2 with N_2 ideal gas molecules, both at a pressure P and temperature T. After removal of the separation, the gas molecules will diffuse and form a homogeneous mixture of $N = N_1 + N_2$ molecules at pressure P and temperature T. Using Eq. (2.53), the entropy of mixing ΔS_M is given by:

$$\Delta S_M = R(N_1 + N_2) \ln (V_1 + V_2) - RN_1 \ln V_1 - RN_2 \ln V_2 \quad [2.57]$$

which can alternatively be written as:

$$\Delta S_M = -R \left[N_1 \ln \frac{V_1}{V} + N_2 \ln \frac{V_2}{V} \right] \qquad\qquad [2.58]$$

As expected the entropy of mixing is always positive. Similar concept has been introduced as a starting point for the theory of liquid mixtures. In terms of interactions, an ideal gas was characterized by a complete absence of repulsive and attractive forces between the molecules. In an ideal mixture, interactions are by definition identical. The intermolecular forces between A and A, A and B, and B and B are

Ideal gas

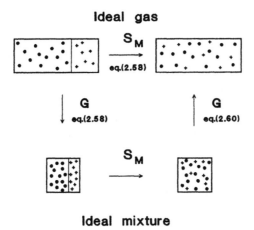

Ideal mixture

Figure 2.10. Succession of transitions to derive the entropy of mixing of an ideal mixture at constant temperature.

all the same. The free energy of mixing of an ideal mixture can be calculated from the succession of steps shown in Figure 2.10. This is done by comparison of two routes: direct mixing in the gas phase $(1 \to 4)$ and mixing in the dense liquid phase $(1 \to 2 \to 3 \to 4)$. We start with a vessel with two compartments at a pressure P and temperature T, with respectively N_1 and N_2 molecules. The volumes V_1 and V_2 are so large that the systems can be treated as ideal gases. If the separation is removed $(1 \to 4)$, the particles mix with an entropy of mixing, given by Eq. (2.58). Another way of reaching this state is by first condensing both separated compartments to liquid density $(1 \to 2)$. Since both types of molecules have identical mutual interactions, this state is obtained with the same pressure P_L. The corresponding change of free energy ΔG depends on the equation of state of the molecules and is given by (constant T):

$$\Delta G = N_1 \int_{P}^{P_L} VdP + N_2 \int_{P}^{P_L} VdP \qquad [2.59]$$

The next step is the removal of the separation and subsequent mixing of the molecules. Since all interactions are the same, this only involves the entropy of mixing ΔS_M. Finally, the mixture is expanded to the volume $V = V_1 + V_2$. Since the molecules have identical interactions, the equation of state of the mixture is the same as of the pure components and the change of the Gibbs free energy is given by:

$$\Delta G = (N_1 + N_2) \int_{P_L}^{P} V dP \qquad\qquad [2.60]$$

Clearly, the net value of the integrals over VdP (Eq. (2.59) + Eq. (2.60)) vanishes. The total change of the Gibbs free energy of the system, over the second route is thus entirely entropic. The entropy of mixing is given by the corresponding expression for ideal gases, Eq. (2.58). Since volumes are proportional to the number of particles ($N_i \propto V_i$), the entropy of mixing of an ideal mixture can be expressed as:

$$\frac{\Delta S_M^{ideal}}{NR} = -[X_1 \ln X_1 + X_2 \ln X_2] \qquad\qquad [2.61]$$

and the Gibbs free energy of mixing of an ideal mixture is simply given by:

$$\Delta G_M^{ideal} = -T\Delta S_M^{ideal} \qquad\qquad [2.62]$$

In this sense, there is no difference with an ideal gas. From the Gibbs free energy of mixing, the chemical potential $\Delta\mu_1$ of molecule A (index 1) in the mixture relative to the pure liquid A can be derived:

$$\Delta\mu_1 = \left(\frac{\partial \Delta G_M}{\partial N_1}\right)_{P,T,N_2} = RT \ln X_1 \qquad\qquad [2.63]$$

Eq. (2.63) is used as an alternative definition of an ideal mixture.

We will now derive an expression for the vapor pressure of an ideal mixture, given the vapor pressures of each pure component, assuming that the vapor behaves as an ideal gas. With Eq. (2.53) and $G = -TS$, one obtains for the chemical potential $\Delta\mu^{vap}$ of a molecule in an ideal gas relative to a state at a reference (partial) pressure P_{ref}:

$$\Delta\mu_1^{vap} = RT \ln \frac{P_1}{P_{ref}} \qquad\qquad [2.64]$$

where P_1 is the partial pressure of component A. Consider a pure liquid A in equilibrium with its vapor at temperature T. The pressure is the vapor pressure $P_1^{vap}(T)$ of molecule A. This will be the reference state. Now consider also a mixture of A and B at the same temperature T in equilibrium. The partial pressure of component A is given by P_1^{mix}. Imagine we transfer (isothermally) a molecule A from the liquid mixture to a pure liquid A. This involves a change in the chemical potential, given by Eq. (2.63). Transferring a molecule A from the vapor phase of the mixture to the vapor phase of pure A involves a change in the chemical potential, given by Eq. (2.64) with $P_{ref} = P_1^{vap}$. As both systems are in equilibrium both chemical potential changes should be equal. One thus obtains:

$$P_1^{mix} = X_1 P_1^{vap}$$ [2.65]

In other words: for an ideal mixture, the partial pressure of each component is equal to the mole fraction in the mixture times the vapor pressure of the pure material at that temperature, provided the vapors behave as an ideal gas. Eq. (2.65) is known as Raoult's law[1,3,8] and was experimentally verified for a number of systems by F. M. Raoult in 1886.

For non ideal mixtures where the interactions between the molecules are non-uniform, there is still some form of ideality when the mixtures are so dilute in one component, say A, that each A molecule is in a uniform environment (of B molecules). Then, the partial pressure of A is also proportional to its mole fraction in the mixture. Due to the interactions, the proportionality constant is no longer 1 as in ideal mixtures:

$$\frac{P_1^{mix}}{P_1^{vap}} = H X_1$$ [2.66]

This relation for dilute mixtures ($X_1 \ll 1$) is known as Henry's law[1,3,8] and dates back to the beginning of the 19th century.

Deviations from ideality are a measure for the interactions between A and B molecules or more precisely: the deviations from the equality of A-A, B-B and A-B interactions. Therefore, vapor pressure measurements can be used to study these interactions. A more frequently used technique to measure polymer-solvent interactions in particular is osmometry, which is from a thermodynamic point of view fully equivalent to vapor pressure measurements as we will now show.

Consider a pure solvent A which is separated from a mixture of B in A by a semi-permeable membrane that only allows solvent molecules to pass. If B is a high molecular weight polymer, this can be achieved by a membrane with very small pinholes. Both systems are in equilibrium with their respective vapor phases, see Figure 2.11. For convenience, we assume B to be non-volatile. As we have seen above, the vapor pressure of the mixture will be lower than that of the pure solvent, which means that initially there will be a pressure difference across the membrane. This will result in a net flow of A molecules from the pure solvent into the mixture, so that the level of the liquid at the mixture side will increase until the excess hydrostatic

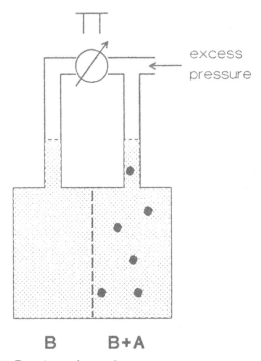

Figure 2.11. Experimental setup for osmometry.

pressure compensates for the lowered vapor pressure and thermodynamic equilibrium is reached. This excess hydrostatic pressure can be
measured and is called the osmotic pressure Π of the mixture. A better
way to do the experiment is to apply excess pressure at the mixture
side so as to keep the levels equal. In the latter set-up the composition
of the mixture does not alter during the experiment. Concluding: the
osmotic pressure of a mixture against the solvent is equal to the
lowering of the vapor pressure of the solvent in the mixture.

In terms of chemical potentials, the equilibrium is reached when
the lowering of μ_1 due to the addition of B (Eq. (2.63) is compensated by
the increase of μ_1 as a result of the applied pressure Π, Eq. (2.9)). This
implies:

$$-RT \ln X_1 = \int_0^{\Pi} V_1 dP \qquad \text{[2.67]}$$

We now assume that the partial molar volume of A is independent of
pressure, which implies that the compressibility of the mixture is
negligible. We also assume that it is independent of composition,
which implies that one may take the molar volume of the pure solvent

for V_1. Eq. (2.67) can then be transformed into an explicit expression for the osmotic pressure Π of an ideal mixture:

$$\Pi = -\frac{RT}{V_1} \ln X_1 \qquad [2.68]$$

This can be further simplified by replacing X_1 by $(1 - X_2)$ and expanding the logarithm to the first term for sufficiently dilute mixtures:

$$\Pi = \frac{RT}{V_1} X_2 \qquad [2.69]$$

One last approximation for dilute mixtures, namely $X_2/V_1 \approx c$, where c is the molar concentration of B leads to Van't Hoff's law:

$$\Pi = cRT \qquad [2.70]$$

With $c = N_2/V$, Eq. (2.70) is similar to the ideal gas law and may be referred to as the ideal mixture osmotic equation of state:

$$\Pi = \frac{N_2 RT}{V} \qquad [2.71]$$

Written with $c = W_2/M_2$ ($c << 1$), where W is the weight concentration and M_2 is the molecular weight of the solute B one obtains:

$$\Pi = \frac{1}{M_2} W_2 RT \qquad [2.72]$$

which shows that osmotic pressure measurements with dilute solutions can be used to determine the molecular weight of the solute.

2.4 REGULAR MIXTURES

2.4.1 DEVIATIONS FROM IDEALITY

It will be no big surprise that ideal mixtures are hard to find. Nevertheless the concept has proven to be, just like the ideal gas, a useful tool in building the theoretical framework for the description of more realistic mixtures. The physical picture of an ideal mixture is one in which the constituent molecules have entirely symmetrical interactions. A-A, B-B and A-B interactions are identical. The A and B molecules are mixed randomly and the entropy of mixing is given by Eq. (2.61), which is essentially the ideal gas entropy of mixing. Molecules in real mixtures do not show identical interactions. As a result, mixing not only involves entropy changes but also energy changes, manifested by heat effects on mixing. The Gibbs free energy of mixing becomes:

$$\Delta G_M = \Delta H_M - T\Delta S_M \qquad\qquad [2.73]$$

where ΔH_M is the heat that is consumed on mixing (at constant temperature and pressure). A positive ΔH_M means endothermic mixing and a positive contribution to ΔG_M. If not compensated by a sufficiently large entropy of mixing, such a combination of molecules will not mix into a homogeneous mixture. For such cases, the heat of mixing is of course an entirely theoretical concept: It is the heat that would be consumed if the molecules were to form a uniform mixture.

2.4.2 REGULAR MIXTURES

In order to describe these effects mathematically, the concept of the so-called regular mixture has been introduced by Hildebrand in 1929. A regular mixture is, by definition, a mixture with an ideal entropy of mixing, given by Eq. (2.61) but a non-zero heat of mixing. The physical picture is a mixture that still mixes randomly although the net effect of breaking A-A and B-B interactions for A-B interactions is a change in the energy of the system. In a mean field approximation, we will derive an explicit expression for ΔH_M. Define the following quantities:

N_1	number of molecules A
N_2	number of molecules B
$N = N_1 + N_2$	total number of molecules
z	number of neighbors of each molecule
E_{11}	energy of an A-A contact
E_{12}	energy of an A-B contact
E_{22}	energy of a B-B contact

Let X_1 and X_2 be the mole fractions of A and B molecules in the mixture. The energy contents H_1 and H_2 of the pure liquids are given by:

$$H_1 = -\frac{1}{2}zN_1E_{11} \qquad\qquad [2.74]$$

$$H_2 = -\frac{1}{2}zN_2E_{22}$$

because each of the N_i molecules is surrounded by z neighbors with a total interaction energy zE_{ii} per pair and hence $zE_{ii}/2$ per molecule.

In the random mixture each molecule is on average surrounded by zX_1 molecules of A and zX_2 molecules of B. The total energy H_{12} of the mixture is therefore given by:

$$H_{12} = -\frac{1}{2}zN_1[X_1E_{11} + X_2E_{12}] - \frac{1}{2}zN_2[X_1E_{21} + X_2E_{22}] \qquad [2.75]$$

Note that with this choice of signs, the attractive energies E that hold the liquid together are counted positive. For the heat of mixing $\Delta H_M = H_{12} - H_1 - H_2$, one thus obtains:

$$\Delta H_M = zNWX_1X_2 \qquad [2.76]$$

with the exchange energy W given by:

$$W = \frac{1}{2}\left(E_{11} + E_{22}\right) - E_{12} \qquad [2.77]$$

where we also used that $E_{12} = E_{21}$. One thus arrives at the following expression for the Gibbs free energy of mixing of a regular mixture:

$$\frac{\Delta G_M}{NRT} = X_1 \ln X_1 + X_2 \ln X_2 + \chi X_1 X_2 \qquad [2.78]$$

It is common practice to write $\Delta G_M/NRT$, which is dimensionless, instead of ΔG_M. We also introduced the so-called interaction parameter χ in anticipation of the extension of this theory to polymeric mixtures in the next section. The concept is somewhat confusing because the interaction parameter of the above discussed model mean field regular mixture is not constant but temperature dependent:

$$\chi = \frac{zW}{RT} \qquad [2.79]$$

2.5 SOLUBILITY PARAMETERS

Molecular interaction energies E in the above derivation can be related to some pure component properties. The method is particularly employed in polymer science but since it is in no way particular to polymers it will be discussed here. Interaction energies E_{ij} related to dispersion forces, such as London-Van der Waals interactions often depend as $\Gamma_i\Gamma_j$ on some molecular property Γ (e.g., the molecular polarizability). In such a case, the E_{ij} can be calculated from the pure component values E_{ii} and E_{jj} according to the geometric mean rule:

$$E_{ij} = (E_{ii}E_{jj})^{1/2} \qquad [2.80]$$

For the exchange energy, W, one then obtains the simple expression:

$$W = \frac{1}{2}(E_{11} + E_{22}) - (E_{11}E_{22})^{1/2} = \frac{1}{2}(E_{11}^{1/2} - E_{22}^{1/2})^2 \qquad [2.81]$$

with the important conclusion that the heat of mixing of such mixtures is always positive (endothermic mixing) or, in the special case, that $E_{11} = E_{22}$, zero.

The internal interaction energy E_{ii} is related to the energy of vaporization E^{vap} of the material. Within the above concept, the removal of one molecule from the liquid to infinity has the net effect of breaking $z/2$ interactions and thus requires an amount of energy $zE_{ii}/2$. If V_m is the molar volume of the material then the energy of vaporization per unit volume is given by:

$$\frac{U_{vap}}{V} = \frac{\frac{1}{2}zE_{ii}}{V_m} \qquad [2.82]$$

The left hand side of Eq. (2.82) is the cohesive energy density (often denoted CED) of the material. For reasons that will become clear below, the square root of the CED is called the solubility parameter (denoted δ). One thus has:

$$\delta_i^2 = \frac{\frac{1}{2}zE_{ii}}{V_m} \qquad [2.83]$$

Substitution into Eq. (2.79) and Eq. (2.81) yields the following simple expression for the interaction parameter in terms of solubility parameters:

$$\chi = \frac{V_m}{RT}(\delta_1 - \delta_2)^2 \qquad [2.84]$$

Eq. (2.84) explains the origin of the 'like dissolves like' principle. If the solubility parameter difference is too large, the positive heat of mixing contribution to the Gibbs free energy of mixing is larger than the negative entropy of mixing contribution. The resulting positive ΔG_M implies that the two liquids will not mix. More generally: the smaller the difference of the solubility parameters, the more likely a homogeneous mixture will be formed, hence: like dissolves like.

According to Eq. (2.82), solubility parameters can be calculated from the energy of vaporization (internal energy) U_{vap} (per mole). This should not be confused with the latent heat of vaporization (enthalpy) H_{vap} which is the energy that is required to bring a mole of liquid to the vapor phase at constant temperature and pressure. The difference between U_{vap} and H_{vap} is the energy of the vapor phase. Assuming that the vapor can be represented by an ideal gas one has:

$$U_{vap} = H_{vap} - RT_{ref} \qquad [2.85]$$

where T_{ref} is the temperature at which H_{vap} has been measured. One thus finds the following relation between the solubility parameter δ

of a liquid with density ρ and molecular weight M (N/V = ρ/M) and the experimental heat of vaporization H_{vap}:

$$\delta = \left(\frac{\rho(H_{vap} - RT_{ref})}{M} \right)^{1/2}$$

[2.86]

This method of measuring is only suitable for low molecular weight substances that can be evaporated. Polymers need very high temperatures to evaporate and will rather disintegrate due to their limited thermal stability.

Empirical relations exist between the solubility parameter and other physical properties, related to energy densities, that can be measured without evaporating the material. Examples are:

$$\delta^2 = \frac{\alpha T}{\kappa}$$

[2.87]

$$\delta^2 = 13.5 \frac{\gamma}{V_m^{1/3}}$$

where α is the thermal expansion coefficient, κ is the isothermal compressibility and γ is the surface tension.

2.5.1 GROUP CONTRIBUTION METHODS

There is a very useful method to calculate the solubility parameter of a molecule on the basis of its chemical structure. The basic assumption is that each fragment (such as a CH_3 group) in a molecule has an interaction with the other fragments which is independent of the location of the fragment in the particular molecule. In the previous derivation, the molar volumes of each constituent were assumed to be equal (denoted V_m). Then it is reasonable to assume that "the amount of interactions" of molecule i with other molecules is weighted by their respective mole fractions. The molar volumes of different molecular fragments will generally not be identical and the above derivation should be adjusted.

In order to keep the argument transparent, let us consider a mixture of two different groups. We will now assume that the amount of cohesive interaction energy of one molecule 1 with molecules 2 is proportional to the volume fraction Φ_2 instead of the mole fraction x_2 as used above. The parameter z will be abandoned and incorporated in the cohesive energy parameter E_{ij}. The total cohesive energy of the mixture is now given by:

$$E = n_1\Phi_1E_{11} + n_2\Phi_2E_{22} + n_1\Phi_2E_{12} + n_2\Phi_1E_{21}$$

[2.88]

Note that we haven't put $E_{12} = E_{21}$. There is another symmetry rule as can be seen from the following argument. $n_1\Phi_2E_{12}$ is the total cohesive energy associated with $1 - 2$ interactions as seen from the n_1 molecules 1. On the other hand, $n_2\Phi_1E_{21}$ is the total cohesive energy associated with $2 - 1$ interactions as seen from the n_2 molecules 2. Both expressions should give the same answer: the total intermolecular cohesive energy. Therefore:

$$n_1\Phi_2E_{12} = n_2\Phi_1E_{21} \qquad [2.89]$$

Substituting the expressions for the volume fractions:

$$\Phi_i = \frac{n_iV_i}{V_{tot}} \qquad [2.90]$$

where:

$$V_{tot} = n_1V_1 + n_2V_2 \qquad [2.91]$$

is the total volume of the mixture, yields the symmetry rule for the intermolecular cohesive energy:

$$\frac{E_{12}}{V_1} = \frac{E_{21}}{V_2} \qquad [2.92]$$

which basically states that the symmetrical variable is not the molar cohesive energy but rather the cohesive energy density. In terms of cohesive energy densities $\tilde{U}_{ij} = E_{ij}/V_i$ and $\tilde{U} = E/V_{tot}$, Eq. (2.88) can now be written as:

$$\tilde{U} = \Phi_1^2\tilde{U}_{11} + \Phi_2^2\tilde{U}_{22} + 2\Phi_1\Phi_2\tilde{U}_{12} \qquad [2.93]$$

Following Scatchard[10] we assume that $\tilde{U}_{12}$ is the geometric mean of $\tilde{U}_{11}$ and $\tilde{U}_{22}$:

$$\tilde{U}_{12} = (\tilde{U}_{11}\tilde{U}_{22})^{1/2} \qquad [2.94]$$

with the same arguments as used for Eq. (2.80). Eq. (2.93) can then be written as:

$$\tilde{U} = (\Phi_1\tilde{U}_{11}^{1/2} + \Phi_2\tilde{U}_{22}^{1/2})^2 \qquad [2.95]$$

This expression is the basis of several important results. First, the heat of mixing per unit volume $\Delta H_M/V = \Phi_1\tilde{U}_{11} + \Phi_2\tilde{U}_{22} - U$ follows after some rearrangement:

$$\frac{\Delta H_M}{V} = \Phi_1\Phi_2(\delta_1 - \delta_2)^2 \qquad [2.96]$$

where we used the definition of the solubility parameter as the square root of the cohesive energy density, i.e. $\delta_i^2 = \tilde{U}_{ii}$. A direct result of Eq. (2.95) is an expression for the solubility parameter of the mixture:

$$\delta = \Phi_1\delta_1 + \Phi_2\delta_2 \qquad [2.97]$$

For an arbitrary number of components, one derives the general expression:

$$\delta = \sum_i \Phi_i\delta_i \qquad [2.98]$$

This shows that the solubility parameter of a mixture of molecules with different solubility parameters is just the volume average of the individual solubility parameters.

However, volume fractions as composition variables are not practical. P. A. Small noted[11] in 1953 that Eq. (2.95) may also be written in terms of total cohesive energies E (using Eq. 2.90) as:

$$(EV)^{1/2} = n_1(E_{11}V_1)^{1/2} + (E_{22}V_2)^{1/2} \qquad [2.99]$$

This shows that $(E_iV_i)^{1/2}$ is an additive property which means that it can be calculated as a sum of contributions of the various groups. More specifically, we may write for the solubility parameter $\delta = (E/V)^{1/2}$ ($=[EV]/V]^{1/2}$) of a molecular soup of an arbitrary number of groups:

$$\delta = \frac{\sum_i F_i}{V} \qquad [2.100]$$

where F_i represents $(E_{ii}V_i)^{1/2}$ and V is the (number averaged) molar volume of the molecular soup. If our soup represents different groups on a single molecule, V is just the molar volume of that molecule. By correlating experimental cohesive energy densities (at 25°C) of a large number of molecules, Small was able to compile a table of the so-called molar attraction constants F of various molecular fragments. Later, new tables have been proposed by others, notably Hoy[12], van Krevelen,[13] and Fedors,[14] but Small's tables are still frequently used in polymer science. Table 2.1 is a list of molar attraction constants according to Small, Hoy and van Krevelen. Fedors employed a slightly different approach than depicted by Eq. (2.100). Instead, he used group contributions e_i and v_i to respectively the cohesive energy and molar volume, such that:

Table 2.1. Molar attraction constants of molecular fragments according to various group contribution schemes.

Group	Comment	Small	Hoy	van Krevelen
-CH₃	methyl	214.00	147.30	205.00
-CH₂-	carbon with 2 protons	133.00	131.50	137.00
>CH-	carbon with 1 proton	28.00	85.99	68.00
>C<	carbon without protons	-93.00	32.03	
=CH₂	double bonded	190.00	126.54	
=CH-	double bonded	111.00	121.53	109.00
>C=	double bonded	19.00	84.51	40.00
CH#C-	triple bonded	285.00		
-C#C-	triple bonded	222.00		
=CH- ar	connected to aromatic group		117.12	
=C- ar	connected to aromatic group		98.12	
>C=O	ketone	275.00	262.96	335.00
-CHO	aldehyde		292.64	
-O- eth	as in ether or OH	70.00	114.98	125.00
-O- epoxy	as in epoxy group		176.20	
-COO-	ester	310.00	326.96	250.00
-CO₃-	carbonate			375.00
-C#N	nitril	410.00	354.56	480.00
-OH	hydroxy		226.00	369.00
-OH ar	hydroxy at aromatic group		170.99	
-N-	tertiary amine		61.08	
-NH-	secondary amine		180.03	
-NH₂	primary amine		226.56	
-N=C=O	isocyanate		358.66	
-ONO₂	nitrate	440.00		
-NO₂	aliphatic nitro compound	440.00		
-F	fluoro		41.33	80.00
-Cl sg	only 1 chlorine	270.00	205.06	230.00
-Cl₂ dn	2 chlorines as in CCl₂	520.00	342.67	
-Cl₃ tr	3 chlorines as in CCl₃	750.00		
-Br	bromine	340.00	257.80	300.00
-Br ar	bromine at aromatic group		205.60	
-I	iodine	425.00		
-S-	sulfide	225.00	209.00	225.00
-SH	thiols	315.00		
-H	hydrogen	90.00		
-<=>- ph	phenyl	735.00		741.00
-<=>- ph	phenylene (o,p,m)	685.00		673.00
Conjug.	additional contribution	25.00	23.26	
Ring 5	additional contribution	110.00	20.90	
Ring 6	additional contribution	100.00	-23.44	
Cis	additional contribution		-7.13	
Trans	additional contribution		13.50	
Ortho	additional contribution		9.69	
Para	additional contribution		40.33	
Meta	additional contribution		6.60	

$$\delta = \left(\frac{\sum\limits_i e_i}{\sum\limits_i v_i} \right)^{1/2} \qquad\qquad [2.101]$$

With this tool, one has an entirely predictive method to calculate the interaction parameter χ in Eq. (2.84). One may thus predict the thermodynamic phase behavior of a mixture on the basis of the chemical structure of the constituents. Of course, nature is not that easy on us and quantitative agreement with experiments is too much to hope for. There are many other factors that contribute to ΔG_M. Nevertheless, the dispersion forces that are described by the solubility parameter concept are always present and can be seen as to form a sort of background interaction to the heat of mixing on top of which the other effects contribute. Indeed, the concept has proven to be useful in predicting whether a certain mixture will form a homogene- ous phase, especially in the paint industry where complex mixtures of solvents and polymers occur. In Chapter 4 we will show that also for polymer blends the concept is valid. In polymer science, the solubility parameter concept with the corresponding group contribu- tion method still is the only truly predictive method for the thermo- dynamic miscibility. That is why there is still active research going on to improve the capabilities of calculating solubility parameters.

A limitation of the group contribution approach is that only those molecular structures can be handled that only contain groups that are covered by the scheme. Recently, a novel approach has been worked out to tackle this problem by J. Bicerano. His topological method will now briefly be discussed.

2.5.2 TOPOLOGICAL CONTRIBUTIONS

Group contribution methods for the prediction of physical properties may be used for those properties that increase linearly with the number of groups involved. Examples of such properties are molecular weight, molar volume, and the quantity $(EV)^{1/2}$ as discussed in the previous section. These are extensive properties.

Group contribution methods are empirical methods. They build on a large set of experimental data in the form of chemical structures and corresponding properties. For the application of group contribu- tion methods, a (limited) number (N_G) of structural groups is defined, such that each chemical structure from the data set can be seen as being built up from these groups. Let n_{ij} be the number of groups with index j (j=1...N_G) in molecule i and let P_i be the experimental value of the desired property of molecule i. Then one writes:

$$P_i = \sum_{j=1}^{N_G} n_{ij}F_j \quad i=1...N_P \qquad\qquad [2.102]$$

where N_P is the number of experimental data and F_i is the empirical group additive contribution of group i to the property P. Equation (2.102) constitutes a set of N_P equations with N_G unknowns (F_j). The statistically best values of F_j can be determined from a linear regression analysis.

With an established set F_j, the property P of a novel structure may be predicted, provided it can be constructed from elements that are contained in the set of N_G structural groups. This latter requirement is an important limitation of group contribution schemes. For a statistically sound incorporation of a structural group in the scheme, a large number of experimental data of molecules containing that particular group is needed. Novel structures have of course a tendency not to be covered by existing schemes.

One way to overcome this difficulty has been developed by Seitz.[16] He observed that many physical properties can be correlated to a limited number of 'fundamental properties': molecular weight of the repeat unit, polymer backbone length, Van der Waals volume of repeat unit, cohesive energy, and a parameter related to the rotational degrees of freedom of the backbone chain. The toughest parameter to estimate is the cohesive energy which also happens to be the most relevant parameter for the thermodynamic miscibility. The problem of the limited number of groups was overcome by Bicerano[16] with a topological technique.

The procedure is as follows. First, the molecule (or polymer repeat unit) is simplified by ignoring all protons. The molecular structure becomes a hydrogen suppressed graph with each remaining (non-hydrogen) atom acting as a vertex and each bond as an edge. Figure 2.12 shows a polymethylmethacrylate (PMMA) repeat unit and its corresponding hydrogen suppressed graph.

The next step is to assign two indices δ and δ^v to each vertex. The so-called connectivity index δ is simply the number of edges emanating from the vertex (i.e., the number of non-hydrogen atoms connected to the atom). The valence connectivity index δ^v is defined by:

$$\delta^v = \frac{\text{Number of non–hydrogen coupled outer shell electrons}}{\text{Number of inner shell electrones } - 1} \qquad [2.103]$$

This index is a somewhat arbitrary measure of the electron accepting power of the vertex. Many electrons in the outer shell give an atom strong electron accepting power but this effect is tempered if there are

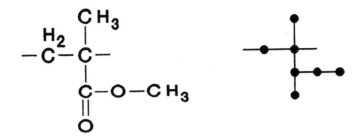

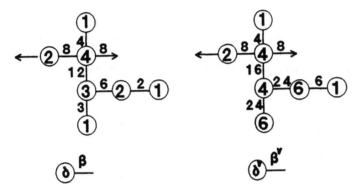

Figure 2.12. PMMA repeat unit, its hydrogen suppressed graph and assignment of atomic and bond indices.

many electrons in the inner shell(s). Also, δ^v is inversely related to the size of the atom.

Two bond indices β and β^v are defined for each edge as the product of the two connectivity indices of the 2 vertices i and j at each side of the edge:

$$\beta = \delta_i \delta_j \qquad\qquad\qquad [2.104]$$

$$\beta^v = \delta_i^v \delta_j^v$$

The assignment of atomic and bond indices to the PMMA repeat unit is also shown in Figure 2.12. All these indices are used to define the so-called zeroth order atomic connectivity indices of the entire molecule or repeat unit:

$$^0\chi = \sum_{\text{vertices}} \frac{1}{\delta^{1/2}} \qquad\qquad\qquad [2.105]$$

$$^{0}\chi^{v} = \sum_{\text{vertices}} \frac{1}{(\delta^{v})^{1/2}}$$

Similarly, the first order (bond) connectivity indices are defined as:

$$^{1}\chi = \sum_{\text{edges}} \frac{1}{\beta^{1/2}} \qquad \qquad [2.106]$$

$$^{1}\chi^{v} = \sum_{\text{edges}} \frac{1}{(\beta^{v})^{1/2}}$$

For the calculation of the bond indices β of a polymeric repeat unit, one of the two bonds connecting the repeat unit with the rest of the molecule is taken into account. One easily sees that the choice of this bond is arbitrary as both bonds will have the same values. For the sample PMMA repeat unit in Figure 2.12 one obtains:

$$^{0}\chi = 5.4916$$
$$^{0}\chi^{v} = 4.5236$$
$$^{1}\chi = 3.1885$$
$$^{1}\chi^{v} = 2.2736$$

Following the above procedure, an arbitrary polymer repeat unit can be characterized with such a set of 4 indices. Linear correlations between a property P and these indices may now be sought:

$$P = a\,{}^{0}\chi + b\,{}^{0}\chi^{v} + c\,{}^{1}\chi + d\,{}^{1}\chi^{v} \qquad \qquad [2.107]$$

by linear regression fits of a, b, c, and d. Preferably these fits should be made to experimental values. Unfortunately, experimental data are not always available and show scattering. Alternatively, the parameters may be fit to group contribution predictions from existing tables. The current method would then offer the advantage of a much greater generality.

For the calculation of solubility parameters, the molar volume and cohesive energy are the relevant parameters. The experimental room temperature molar volumes V_m of a set of 152 polymers were fitted to Eq. (2.107). A strong correlation (correlation coefficient 0.9888) of V_m with the $^{1}\chi^{v}$ was found:

$$V_m = 33.585960\,{}^{1}\chi^{v} \qquad \qquad [2.108]$$

In order to incorporate silicon containing polymers in the correlations it appeared to be necessary to make some special provisions. For the calculation of the bond indices, all Si atoms must be replaced by carbon atoms and a correction term for the number N_{Si} of silicon atoms must be made. The expression:

$$V_m = 33.585960 \, ^1\chi^v + 26.518075 N_{Si} \qquad [2.109]$$

has a 8.6 cc/mole standard deviation and a correlation coefficient of 0.9931. Part of the deviations are of course related to experimental error margins. By searching for systematic deviations in the correlation related to the presence of specific groups, an even more precise correlation was found:

$$V_m = 3.642770 \, ^0\chi + 9.798697 \, ^0\chi^v - 8.542819 \, ^1\chi +$$

$$+ 21.693912 \, ^1\chi^v + 0.978655 \, N_{MV} \qquad [2.110]$$

where N_{MV} follows from the numbers N_X of particular groups X in the repeat unit according to:

$$N_{MV} = 24 N_{Si} - 18 N_{-S-} - 5 N_{sulfone} - 7 N_{Cl} - 16 N_{Br} +$$

$$+ 2 N_{backbone\ ester} + 3 N_{ether} + 5 N_{carbonate} + 5 N_{C=C} -$$

$$- 11 N_{cyc} - 7 (N_{fused} - 1) \qquad [2.111]$$

where N_{cyc} is the number of rings with no double bonds along the edges and N_{fused} is the number of ring structures that share a side. Eq. (2.110) has a standard deviation of 3.2 cc/mol and a correlation coefficient of 0.9989.

Experimental data of cohesive energy densities show much more scatter than those of the molar volumes. This hinders the fit to the data. This problem was circumvented by fitting the coefficients of the respective connectivity indices to the group contribution predictions. This limits the input data set to those structures that are covered by existing group contribution schemes. By nature of the topological method, predictions are not limited to these structures anymore. This enhanced generality is the main advantage of this method. Fits of the cohesive energy E_{coh} were published, using the methods of Fedors[14] (E_{coh}^{Fed}) and van Krevelen[13] (E_{coh}^{vK}) as well as to the dispersion component F_d of the molar attraction constant (see Eq.(2.100)). Where a simple correlation of the molar volume V_m with $^1\chi^v$ already had a correlation coefficient of 0.988, cohesive energies turned out to be quite a lot more difficult to fit. For E_{coh}^{Fed} the strongest correlation was found with $^1\chi$ with a correlation coefficient of "only" 0.9487. A significant improvement was obtained by using a correction term (just like in Eq. (2.110)):

$$E_{coh}^{Fed} = 9882.5 \, ^1\chi + 358.7 (6 N_{atomic} + 5 N_{group}) \qquad [2.112]$$

with:

$$N_{atomic} = 4N_{-S-} + 12N_{sulfone} - N_F + 3N_{Cl} + 5N_{Br} + 7N_{cyanide} \qquad [2.113]$$

and:

$$N_{group} = 12N_{hydroxyl} + 12N_{amide} + 12N_{non\text{-}amide\ -(NH)-} - N_{alkyl\ ether\ -O-} -$$

$$- N_{C=C} + 4N_{non\text{-}amide\ -(C=O)-\ next\ to\ N} +$$

$$+ 7N_{-(C=O)-\ in\ carboxylic\ acid,\ ketone\ or\ aldehyde} +$$

$$+ 2N_{other\ -(C=O)-} + 4N_{N\ atoms\ in\ six\text{-}membered\ aromatic\ rings} \qquad [2.114]$$

This expression has a correlation coefficient of 0.9974 and a standard deviation of 3.9% of the average E_{coh} of 59192 J/mol.

Similarly, a correlation is given for the van Krevelen/Hoftyzer group contribution predictions:

$$E_{coh}^{vK} = 10570.9(^0\chi^v - {^0\chi}) + 9072.8(2^1\chi - {^1\chi^v}) + 1018.2N_{VKH} \qquad [2.115]$$

where all Silicon atoms must be replaced by carbon for the evaluation of Eq. (2.115). The correction term is given by:

$$N_{VKH} = N_{Si} + 3N_{-S-} + 36N_{sulfone} + 4N_{Cl} + 2N_{Br} + 12N_{cyanide} + 8N_{ketone} +$$

$$+ 16N_{non\text{-}amide\ C=O\ next\ to\ N} + 33N_{HB} - 4N_{cyc} + 19N_{anhydride} +$$

$$+ 2N_{N\ with\ \delta=2,\ but\ not\ adjacent\ C=O,\ and\ not\ in\ six\text{-}membered\ aromatic\ ring} +$$

$$+ 7N_{N\ in\ six\text{-}membered\ aromatic\ ring} + 20N_{carboxylic\ acid} +$$

$$+ \sum (4 - N_{row})_{substituents\ with\ \delta=1\ attached\ to\ aromatic\ ring\ in\ backbone} \qquad [2.116]$$

N_{row} is the row in the periodic table in which the atom represented by the vertex with $\delta = 1$ is located (e.g., methyl and fluorine substituents contribute $N_{row} = 1$, chlorine contributes $N_{row} = 2$). For more details we refer to the book.[16] The above correlation has a correlation coefficient of 0.9988 and a standard deviation of 3% of the average E_{coh} value of 55384 J/mol.

Finally, a correlation was given for the dispersion component F_d of the molar attraction constant F (units: $J^{0.5}\ cm^{1.5}\ mole^{-1}$) using the group contribution scheme of van Krevelen. A very good fit (correlation coefficient: 0.9997, standard deviation: 30 $J^{0.5}\ cm^{1.5}\ mole^{-1}$) is obtained by:

$$F_d = 97.95[-{^0\chi} + 2(^0\chi + {^1\chi} + {^1\chi^v})] + 134.61N_{F_d} \qquad [2.117]$$

with:

Table 2.2: Solubility parameters calculated using the three topological methods and the group contribution schemes of Small and Hoy. For the calculation of densities and molar volumes, Eq. (2.110) has been employed.

	E_{coh}^{Fed}	E_{coh}^{vK}	F_d	Small	Hoy
Polyethene	17.5	16.8	16.5	16.9	16.7
Polypropene	16.8	16.1	15.7	15.7	15.2
Polyisobutene	16.0	15.4	15.0	14.6	14.3
Poly(1,4-butadiene)	17.4	17.7	15.9	16.9	17.5
Polyisoprene	16.9	17.2	15.6	16.3	16.5
Polystyrene	20.1	19.5	18.1	18.9	
Polyvinylchloride	21.2	19.6	17.9	19.5	19.1

$$N_{F_d} = N_{Si} - N_{Br} - N_{cyc} \qquad\qquad [2.118]$$

Solubility parameters can be calculated according to:

$$\delta = \left(\frac{E_{coh}}{V_m}\right)^{1/2} \qquad\qquad [2.119]$$

or:

$$\delta_d = \frac{F_d}{V_m} \qquad\qquad [2.120]$$

Table 2.2 gives a comparison of the calculated solubility parameters of a number of (not too polar) polymers using the three topological correlations and the group contribution schemes of Small[11] and Hoy[12].

The variation in the predictions gives an idea of the absolute accuracy of the method.

2.6 LATTICE THEORIES

The derivations in the above discussions were made without reference to any specific structure of the mixture. The only parameter that was related to the liquid structure is the number of neighbors z. This parameter usually enters the discussion as the so-called lattice coordination number. The introduction of a hypothetical lattice on which the molecules are placed enables explicit counting the number of configurations or contact points as the total number of lattice points, i.e., possible positions, is finite. Often, this finite number of possibilities is the only motive for using a lattice. The specific structure of the lattice, for example that turning left 3 times on a cubic lattice means coming back to the starting point, is not used. In the next chapter, we will discuss counting polymer configurations and here the specific lattice structure does play a significant role.

We will now derive some of the previous results, using a lattice model, or phrased better: using a finite number of possible locations, illustrated by configurations on a lattice.

2.6.1 ENTROPY OF MIXING

Imagine a system with n_1 particles that can be placed on any of N_1 possible locations. We shall first consider an ideal gas which means that there is no interaction between the particles and each location can be occupied by any number of particles. The total number of configurations Ω is given by:

$$\Omega = N_1^{n_1} \qquad [2.121]$$

since each location has n_1 possible occupation levels. According to the results of statistical mechanics the entropy S_1 is given by:

$$\frac{S}{R} = \ln \Omega = n_1 \ln N_1 \qquad [2.122]$$

A similar expression holds for an additional system with n_2 different particles on N_2 locations. The total entropy of these two (separated) systems is:

$$\frac{S_1 + S_2}{R} = n_1 \ln N_1 + n_2 \ln N_2 \qquad [2.123]$$

Now we mix the two systems to form one system with $n_1 + n_2$ particles on $N = N_1 + N_2$ possible locations. The entropy S_N of the mixture is given by:

$$\frac{S_N}{R} = (n_1 + n_2) \ln N \qquad [2.124]$$

and the entropy of mixing $\Delta S_M = S_N - (S_1 + S_2)$ follows:

$$\frac{\Delta S_M}{R} = -\left[n_1 \ln \frac{N_1}{N} + n_2 \ln \frac{N_2}{N} \right] \qquad [2.125]$$

With the additional assumption that the number of available locations is proportional to the number of particles the arguments of the logarithmic terms may be replaced by mole fractions and we obtain:

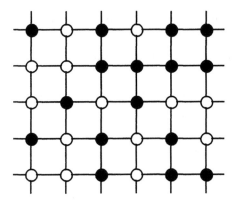

Figure 2.13. Configuration of the mixture on a two-dimensional cubic lattice.

$$\frac{\Delta S_M}{R} = -[n_1 \ln X_1 + n_2 \ln X_2] \qquad [2.126]$$

which is indeed the expression for the entropy of mixing of an ideal mixture.

Now consider the physically more realistic representation of a liquid as a system with N particles that may occupy N locations with exactly 1 particle per location. The pure liquids are just fully occupied systems and contain no additional entropy: $S_1 = S_2 = 0$. The mixture, however, consist of n_1 particles of label 1 and n_2 particles of label 2 which form distinctive configurations (fill patterns) on the $N = n_1 + n_2$ locations. Figure 2.13 shows one such particular configuration on a cubic lattice. The total number of distinctive configurations can be found as follows: Start filling the lattice with particles labeled 1. The first particle has N possible locations, the second, one location now being occupied, has N-1 possibilities. Particle number n_1 has $N-n_1+1$ possibilities. The total number of configurations thus appears to be given by $N(N-1)(N-2)\ldots N-n_1+1 = N!/(N-n_1)! = N!/n_2!$. This, however, is overcounting. The label 1 particles are indistinguishable and one should count particle 1 on position X, particle 2 on position Y and particle 1 on position Y, particle 2 on position X as one. So we have to divide by the total number of ways to divide the n_1 particles over the n_1 positions which is given by $n_1!$. Once the label 1 particles have been placed, the positions of the label 2 particles are fixed and filling up the lattice does not contribute any more to the total number of configurations Ω which is thus given by:

$$\Omega = \frac{N!}{n_1! n_2!} \qquad [2.127]$$

As $S_1 = S_2 = 0$, the entropy of mixing ΔS_M is given by:

$$\frac{\Delta S_M}{R} = \ln \Omega = \ln N! - \ln n_1! - \ln n_2! \qquad [2.128]$$

This may be worked out using Stirling's expression for large M:

$$\ln M! \approx M \ln M - M \qquad [2.129]$$

and leads again to the expression for the entropy of mixing of an ideal mixture:

$$\frac{\Delta S_M}{R} = -[n_1 \ln X_1 + n_2 \ln X_2] \qquad [2.130]$$

The last derivation clearly shows the advantages of a lattice theory, or rather a finite number of sites, in the use of the rules of combinatorial statistics and such fine tools as Stirling's approximation for $\ln M!$. All sorts of refinements can now easily be incorporated.

2.6.2 HEAT OF MIXING

The derivation of the heat of mixing ΔH_M of a regular mixture in a lattice model is not different from the one discussed in section 2.4. The parameter z is now the lattice coordination number. As such it also is the number of neighbor molecules as in the previous derivation.

REFERENCES

1.   C. J. Adkins, *Equilibrium Thermodynamics*, Cambridge University Press, Cambridge, 1985.
2.   W. T. Grandy, *Foundations of Statistical Mechanics. Vol. 1: Equilibrium theory*, D. Reidel Publishing Company, Dordrecht (Netherlands) 1985.
3.   S. M. Walas, *Phase equilibria in chemical engineering*, Butterworth Publishers, Stoneham, 1985.
4.   J. W. Gibbs, *Transactions of the Connecticut Academy III*, p. 108-248, 343-524 (1875-1878).
5.   J. W. Gibbs, *The collected works of J. Williard Gibbs, Vol. 1: Thermodynamics*, Yale University Press, 1948.
6.   I. C. Sanchez in *Polymer compatibility and incompatibility, theory and practices*, K. Šolc ed., Harwood Academic Publishers, London, 1982.
7.   E. Fermi, *Thermodynamics*, Dover Publications, New York, 1956.
8.   W. J. Moore, *Physical Chemistry*, Prentice Hall, Englewood Cliffs, 1972.
9.   D. M. Koenhen and C. A. Smolders, *J. Appl. Pol. Science*, **19**, 1163 (1975).
10. G. Scatchard, *Chemical Reviews*, **8**, 311 (1931).
11. P. A. Small, *J. Appl Chem*. **3**, 71 (1956).
12. K. L. Hoy, *J. Paint Tech.*, **42**, 79 (1970).
13. D.W. van Krevelen, *Properties of Polymers*, Elsevier, Amsterdam, 1990.
14. R.F. Fedors, *Polymer Eng. Sci.*, **14**, 147 (1974).
16. J. Bicerano, *Prediction of Polymer Properties*, Marcel Dekker, New York, 1993.

Chapter 3

BASIC THERMODYNAMICS OF POLYMERIC MIXTURES

3.1 POLYMERS

3.1.1 INTRODUCTION

Polymers are made up of a large number (Greek: poly) of monomers reacted together by some sort of repetitive chemical reaction into a long chain of covalently bonded chemical groups. Polymers are characterized within the more general concept of macromolecules by the presence of a clear repetitive element. Although, natural polymers do exist (e.g., natural rubber), most polymers are synthesized by polymerization of monomers from the petrochemical industry. Several types of polymers can be distinguished,[1] depending on the structure of the repetitive element. Figure 3.1 illustrates some important types of polymer.

A-A-A-A-A-A-A-A-A-A-A-A-A-....   : Homopolymer A

A-B-A-B-A-B-A-B-A-B-A-B-A-....   : Alternating AB Copolymer

A-A-B-A-B-B-A-B-B-B-A-A-B-....   : Random AB Copolymer

A-A-A-A-....-A-B-....-B-B-B-B-B   : (Di) Block Copolymer AB

A-A-A-A-..A-B-B-..-B-A..-A-A-A   : (Tri) Block Copolymer ABA

A-A-A-A-A-A-A-A-A-A-A-A-A-....   : Grafted Copolymer (B to A)
      B   B
      B   B

Figure 3.1. Various types of polymer structure.

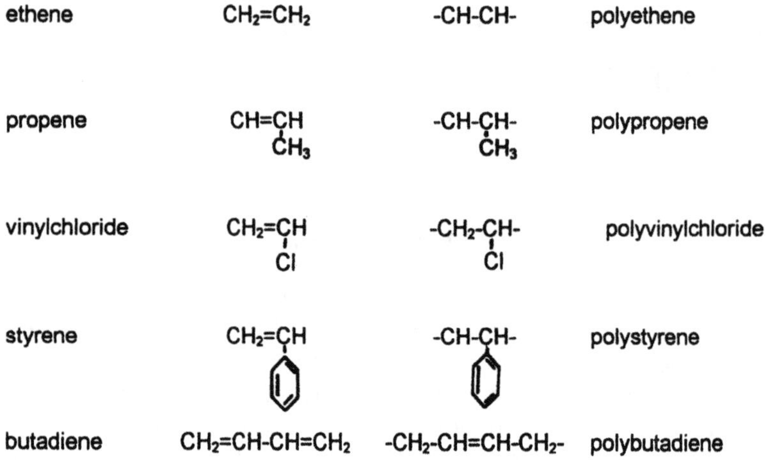

ethene	$CH_2=CH_2$	-CH-CH-	polyethene
propene	CH=CH CH_3	-CH-CH- CH_3	polypropene
vinylchloride	$CH_2=CH$ Cl	$-CH_2-CH-$ Cl	polyvinylchloride
styrene	$CH_2=CH$	-CH-CH-	polystyrene
butadiene	$CH_2=CH-CH=CH_2$	$-CH_2-CH=CH-CH_2-$	polybutadiene

Figure 3.2. Chemical structure of monomers and corresponding repeat units in some well known polymers.

Figure 3.2 shows some well-known monomers with corresponding repeat units. The distinction between homopolymer and alternating copolymer reflects the chemical synthesis (from monomers A and B) rather than the physical structure. For example, polystyrene can just as well be seen as a homopolymer of CH_2-CH-Ø as an alternating copolymer of CH_2 and CH-Ø.

The most important characteristics of a polymer are the chemical structure and the degree of polymerization (number of monomers) D_p. The degree of polymerization is usually expressed in the molecular weight M of the polymer:

$$D_p = \frac{M}{M_r}$$

[3.1]

where M_r is the molecular weight of the repeat unit. For polyethene (the name polyethylene is obsolete) $M_r = 28$ g/mol (the actual smallest repeat unit of polyethene is CH_2, therefore polythene is sometimes also referred to as polymethene (polymethylene), taking C_2H_4 as the repeat unit reflects the type of the monomer used. A polyethene molecule with a molecular weight of 10^6 g/mol thus requires about 36,000 consecutive addition reactions of ethene and consists of 72,000

CH_2 groups. Such a molecular chain would be 0.01 mm long when fully stretched! With the CH_2 groups scaled to the size of a pearl one would have a 300 meter long necklace!

As discussed below polymer chains are usually coiled up. The $M = 10^6$ polyethene molecule from the above example forms coils in the melt or in solution with a typical radius of 50 nm. Let us, for a moment, replace the coil by a sphere with $R_c = 50$ nm radius and filled with the 72,000 CH_2 groups. One finds that only about 0.2% of the sphere volume is filled with polymer segments. The rest is occupied by solvent molecules or, in a melt, with other polymers. Polyethene is an extreme case because it is such a "thin" molecule and $M = 10^6$ g/mol is also rather extreme, though certainly not unrealistic. Polystyrene with $M = 10^5$ g/mol has $R_c = 9$ nm and 5% polymer occupancy.

The conclusion remains that a high molecular weight polymer coil forms a rarefied structure. Single chain statistics will be discussed in some more detail in Section 3.2.

3.1.2 MOLECULAR WEIGHT DISTRIBUTIONS

It is important to realize that it is virtually impossible to synthesize a high molecular weight polymer with an exactly determined molecular weight. One always will have a distribution of molecular weights. Some polymerization processes (such as anionic polymerization) produce by nature a narrow molecular weight distribution. Others (such as radical polymerization) always produce broad distributions. We will now discuss these distributions in some more detail.

Consider a total of N polymer molecules, characterized by a distribution function $X(M)$ where X_i is the mole fraction of polymers with molecular weight M_i. The number average molecular weight M_n is given by:

$$M_n = \sum_i M_i X_i \qquad [3.2]$$

The weight fraction W_i of polymer with molecular weight M_i is given by:

$$W_i = \frac{M_i X_i}{\sum_i M_i X_i} = \frac{M_i}{M_n} X_i \qquad [3.3]$$

And therefore, the weight averaged molecular weight M_w of the distribution is given by:

$$M_w = \sum_i M_i W_i = \frac{\sum_i M_i^2 X_i}{\sum_i M_i X_i} = \frac{\sum_i M_i^2 X_i}{M_n} \qquad [3.4]$$

Higher moments of the distribution can also be defined. Particularly the so-called z-averaged molecular weight M_z is frequently encountered:

$$M_z = \frac{\sum_i M_i^2 W_i}{\sum_i M_i W_i} = \frac{\sum_i M_i^2 W_i}{M_w} \qquad [3.5]$$

With experimental techniques that are sensitive to the number of molecules, such as osmometry, vapor pressure, and melting point depression measurements, one determines the number average molecular weight M_n. With techniques that are sensitive to the size of the molecule, such as all scattering techniques (light, neutrons, x-rays) one determines the weight average molecular weight M_w. The entire distribution can be obtained from ultracentrifugation and gel permeation chromatography (GPC).[1,2]

The width of the molecular weight distribution is often characterized by the so-called heterogeneity index M_w/M_n. For well prepared anionic polymerization processes M_w/M_n may be less than 1.05. Step-growth polymerization[1,2] would give $M_w/M_n = 2$, but in practice larger values are found.[1,2]

3.2 SINGLE CHAIN STATISTICS

3.2.1 THE IDEAL CHAIN

Synthetic polymers are made by polymerizing relatively small monomer molecules into a long chain of connected monomeric groups. We will restrict ourselves to chains with a high degree of flexibility and thus exclude polymers with a backbone or side groups that are so stiff that liquid crystalline mesophases can be formed.[3] The classical example of a flexible polymer chain is polyethene, which is schematically shown in Figure 3.3. Each covalent C-C bond is free to rotate but the steric hindrances of the H groups lead to three preferred torsion angles (0 and $\pm120^\circ$ where the potential energy is minimal). The corresponding chain conformations are respectively called trans, gauche+ and gauche−. This inherent chain flexibility leads to an enormous number of possible configurations, which can only be treated by statistical techniques.

Figure 3.3. Polyethene chain.

Having seen the concept of an ideal solution as a starting point for the theory of mixtures, it is not surprising that the ideal chain has been invented as a starting point for the theory of chain statistics. An ideal chain of segments is a chain with zero interaction between segments that are sufficiently far apart. "Sufficiently far" sets the scale. The chain is ideal, provided it is "sufficiently long". As a consequence of this definition the chain conformations of an ideal chain can be described as random walks, where each step is associated with one (or more) bonds, see Figure 3.4. The fact that these suffi- ciently far apart chain segments do not interact is equivalent to a limited memory of the random walker, allowing him to return to locations that were visited previously but "sufficiently" long time ago. An important characteristic of a chain is the average end-to-end distance R_{ee}. For a random walk with N steps of length L. The root mean square (RMS) averaged $<R_{ee}^2>$ of R_{ee} can easily be calculated:[1,2]

$$<R_{ee}^2>^{1/2} = (NL^2)^{1/2} = N^{1/2}L \qquad [3.6]$$

In the chain model, referred to as the "freely jointed chain", the chain configuration is modeled as the trajectory of a Brownian particle (with equal steps of, e.g., one carbon-carbon bond). The RMS end-to- end distance of such a chain is given by Eq. (3.6). Even if there are preferred angles, the chain dimensions scale with the square root of the number of bonds. Taking the polyethene chain again, one may take

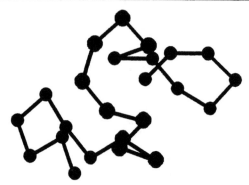

Figure 3.4. Random walk.

the fixed carbon-carbon valence angle (109.5°) into account to arrive at:[4]

$$<R_{ee}^2> = NL^2 \frac{1 - \cos \Theta}{1 + \cos \Theta}$$

[3.7]

For $\Theta = 109.5°$ this implies an expansion of the coil's end-to-end distance by a factor of about 1.4 with respect to the freely jointed chain. The effect of the three preferred torsion angles is similar.[1,2] Side groups will have an effect on the ease of bond rotations and therefore on the dimension of the chain. These effects of local steric hindrances on chain backbone bond rotations are difficult to calculate, although modern molecular modelling software makes such calculations feasible. The steric hindrances are described by the empirical parameter σ, defined by[1]:

$$<r^2>_0 = \sigma^2 NL^2 \frac{1 - \cos \Theta}{1 + \cos \Theta}$$

[3.8]

or alternatively by the characteristic ratio C_∞, defined by:

$$<r^2>_0 = C_\infty NL^2$$

[3.9]

By nature $\sigma > 1$. Equation 3.9 can also be written as:

$$<r^2>_0 = Nb_k^2$$

[3.10]

where b_k is the so-called Kuhn statistical length of the polymer. The polymer chain with local steric hindrances can effectively be replaced by an N step fully random walk chain with bond length $b_k = LC_\infty^{1/2}$.

The quantity $<r^2>_0$ is known as the unperturbed dimension (=end-to-end distance) of the chain. Unperturbed in the sense that it is a characteristic of the chemical structure of the chain bonds.

Interactions with other molecules are neglected. Also, the effect of long range steric interactions are not taken into account: the chain is ideal.

The polymer chain can form a large number of different configurations and there is significant entropy associated with this configurational freedom. The probability per unit volume of finding the end segment at a point at the vector $\vec{r_e}$ from the origin is given by:

$$P(\vec{r_e}) = \left(\frac{1}{\pi^{1/2}r_m}\right)^3 e^{-\left(\frac{r_e}{r_m}\right)^2} \qquad [3.11]$$

with:

$$r_m = \left(\frac{2}{3}<r^2>_0\right)^{1/2} \qquad [3.12]$$

The probability per unit length of finding the end segment at a distance r_e from the first segment is found by integrating Eq. (3.11) over a shell with radius r_e:

$$P(r_e) = \left(\frac{1}{\pi^{1/2}r_m}\right)^3 4\pi r_e^2 e^{-\left(\frac{r_e}{r_m}\right)^2} \qquad [3.13]$$

This probability distribution has a maximum at $r_e = r_m$, while the root mean square average distance is $(3r_m/2)^{1/2}$, in agreement with Eq. (3.12). Eq. (3.13) also shows another important characteristic of an ideal chain: it has a gaussian segment density distribution.

Experimentally, polymer end-to-end distances cannot directly be measured. The most appropriate technique to measure the size of a polymer coil is light or neutron scattering. These techniques yield the so-called radius of gyration, which is the root mean square average distance of the polymer segments from the center of gravity $<s^2>_0$. For gaussian distributions, there is a simple relation between the two measures:

$$<r^2>_0 = 6<s^2>_0 \qquad [3.14]$$

If Ω_T is the total number of possible chain configurations then $\Omega_T P(\vec{r_e})$ is the total number of configurations of a chain with end-to-end vector $\vec{r_e}$. The entropy of such a chain is thus given by:

$$S(\vec{r_e}) = k \ln \Omega_T P(\vec{r_e}) = \text{Const} - k \frac{r_e^2}{r_m^2} \qquad [3.15]$$

Assuming that the enthalpic energy H is constant, the free energy is given by $G(r_e) = H - TS(r_e)$. This implies that the force F needed to increase r_e is given by:

$$F = -\frac{\partial G}{\partial r_e} = -2kT\frac{1}{r_m^2}r_e \qquad\qquad [3.16]$$

per chain. Apparently such a chain satisfies Hooke's law: the force is proportional to the elongation. An elastic band is made of a crosslinked network of such entropic springs. Experimentally one finds that the total volume change of a rubber as a result of strain is extremely small (Poisson's ratio[1] ≈ 0.5): strain in one direction is compensated by contraction in other directions. This implies that the average intermolecular distances are constant (contrary to e.g. metals) which justifies the above assumption of constant enthalpy. Indeed, one observes experimentally[5] that the spring constant of an elastic band is proportional to T as in Eq. (3.16). Another property of a material consisting of entropic springs is that it will increase in temperature when it is adiabatically stretched and cooled down when adiabatically relaxed. This is a result of the second law of thermodynamics and entirely similar to what happens in adiabatic compression and expansion of an ideal gas. In an ideal gas, enthalpic interactions are also constant (zero). Anyone in the possession of two hands, a household elastic band, and a temperature sensitive upper lip may verify.

It is often desirable to be able to count the number of configurations and here is where the lattice model is useful again. An ideal chain is represented as an n step random walk on a lattice. Each lattice point has z neighbors, where z is the coordination number of the lattice. For a two-dimensional cubic lattice z = 4, for a three-dimensional close packed lattice z = 12. After placement of the first segment, each of the remaining n–1 segments has z possibilities. The total number of configurations Ω_{conf}, available to the ideal lattice chain is thus given by:

$$\Omega_{conf} = z^{n-1} \qquad\qquad [3.17]$$

As an easy refinement, one may exclude immediate self-reversal of the chain: segment i+2 cannot be placed on segment i. This leaves z possibilities for the second segment and z–1 for each of the n–2 subsequent segments and thus:

$$\Omega_{conf} = z(z - 1)^{n-2} \qquad\qquad [3.18]$$

3.2.2 REAL CHAINS

The most important shortcoming of the concept of an ideal chain is the neglect of long range interactions. Apart from the local constraints which basically only affect the statistical length, there are no restrictions on the ideal chain configuration. In particular, chain segments are allowed to overlap. Real chains have a finite molecular volume and

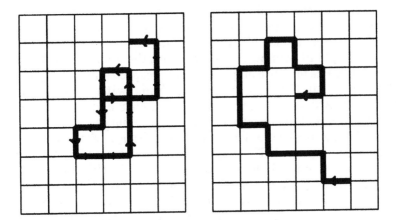

Figure 3.5. Random walk and self-avoiding walk on a two-dimensional lattice.

will exclude overlapping configurations. On a lattice, this can easily be implemented by allowing empty sites for each subsequent segment only. Such configurations are known as Self Avoiding Walks (SAWs). Figure 3.5 shows a random walk and a self avoiding walk on a two-dimensional cubic lattice. This intramolecular excluded volume effect leads to an expansion of the chain with respect to the corresponding ideal chain, since the excluded volume effect is proportional to the segment density which is highest at the center of the chain.

Calculations of chain dimensions and number of configurations were rather trivial for an ideal chain but are much more complex for SAWs. In fact, there is no analytic solution to the problem. Scaling theories provide some answers.[6,7] For example, the total number Ω_{SAW} of SAWs of n steps has the asymptotic form:

$$\Omega_{SAW} \propto z'^n n^{\gamma-1} \tag{3.19}$$

where z′ is somewhat smaller than the coordination number of the lattice (for a 3D cubic lattice $z = 6$ and $z' \approx 4.68$). The attractiveness of scaling theories is that exponents such as γ are universal and depend on the dimension of space only. For all lattices $\gamma \approx 4/3$ in two dimensions and $\gamma \approx 7/6$ in three dimensions. The end-to-end distance of a SAW also scales:

$$R_{ee} \propto r^\nu \tag{3.20}$$

where ν is another universal exponent ($\nu \approx 3/4$ for $D = 2$ and $\nu \approx 3/5$ for $D = 3$).

The numerical values of the exponents can be calculated from exact enumeration (on a lattice) or Monte Carlo methods. Exact enumeration methods only work for small chains. One easily verifies with a simple PC computer program that the number of SAWs on a 2D cubic lattice for 1 to 6 steps amount to respectively 1, 4, 12, 36, 100, 284, and 780. For 24 steps one finds with some patience 46,146,397,316 and then the numbers explode and one soon has to resort to Monte Carlo or other sampling techniques.

There is an elegant derivation, due to Flory,[2] of the approximate value of the exponent. The polymer coil is represented by a D-dimensional sphere (line, circle, sphere for D = 1,2,3) with radius R filled with n polymer segments with mutual repulsive interaction. The concentration of segments $c \propto n/R^D$ and the number of (repulsive) pair interactions $\propto c^2$. Integrating over the total volume of the sphere, one obtains for the total repulsive energy G_{rep}:

$$G_{rep} \propto \frac{n^2}{R^D} \qquad [3.21]$$

The scaling behavior of the elastic free energy G_{el} (entropy) is derived from the ideal chain results; Eqs (3.15), (3.12), and (3.9):

$$G_{el} \propto \frac{R^2}{n} \qquad [3.22]$$

The total free energy G_{tot} is the sum of repulsive and elastic energy and has the following functional dependence on n and R:

$$G_{tot} = A_1 \frac{n^2}{R^D} + A_2 \frac{R^2}{n} \qquad [3.23]$$

with A_1 and A_2 constants. The equilibrium radius corresponds to minimum total free energy, which can be found from $dG_{tot}/dR = 0$. This leads to:

$$R \propto n^{\frac{3}{D+2}} \qquad [3.24]$$

and thus:

$$v = \frac{3}{D+2} \qquad [3.25]$$

This crudely derived expression for v turned out to be remarkably accurate. Values for v that were obtained from computer experiments agree with Eq. (3.25) to within less than a percent. For D = 1, one obtains v = 1, which is correct: a one-dimensional self-avoiding walk is just a straight line with length n. In two dimensions $\nu = 0.75$ while in three dimensions $\nu = 0.6$. As expected, the effect of excluded volume is less in three dimensions. In fact, for d = 4 we obtain $\nu = 0.5$ which is equal to the result of a random walk. We conclude that in four dimensions, "real" polymer chains behave as if they were ideal. This is a rigorous result:[6] in four dimensions, the local concentration of segments in an ideal chain is so low that excluded volume effects become negligible. If we had just lived in one more dimension, than many scaling theoreticians would be unemployed! The accuracy of Eq. (3.25) for D < 5 is believed (by scaling theoreticians!) to be a result of a fortuitous cancellation of errors in the derivation.

3.2.3 A REAL CHAIN IN A SOLVENT

In the above section we have considered one polymer chain in isolation and also neglected attractive forces between segments (apart from the covalent bonds that keep the segments together). Whatever the chemical structure of the polymer, there will always be attractive London type forces between the segments. In this case, configurations with many segment-segment contacts will be energetically more favorable than configurations with fewer contacts. Consequently, the polymer coil will attain a more condensed configuration than in the absence of such attractive interactions. Of course, this goes at the expense of entropy. At equilibrium, the free energy is at a minimum.

Now let's assume that the chain is not in a vacuum but immersed in a solvent, which serves as a background of small molecules with London type dispersive interactions between themselves and with polymer segments. The effect of the solvent on the polymer chain configuration depends on the difference between polymer-polymer, polymer-solvent and solvent-solvent interactions. In the simplest case that all interactions are identical, the coil will not have a preference for more or fewer segment-segment contacts and it will behave as a regular self avoiding walk. Such a solvent is called an athermal solvent as there will be no heat of mixing.

If, on the other hand, polymer-solvent contacts are energetically more favorable than polymer-polymer contacts, the coil will tend to promote such contacts by expanding. Again this goes at the expense of configurational entropy until equilibrium is reached. In the opposite case that polymer-solvent contacts are less favorable (which includes the chain in vacuo), the polymer coil will contract to a more dense configuration. A situation may occur where the repulsive intra-chain excluded volume interactions are exactly compensated by these effectively attractive segment-segment interactions. Since this involves cancellation of enthalpic and entropic contributions to the free energy this will generally only be possible at one particular temperature. This particular state (combination of solvent and temperature) is called the theta (Θ) state. In such a case the "real" chain will show ideal chain behavior. Of course, the chain cannot take "true" random walk configurations just like gas molecules at the Boyle temperature will not truly overlap. However, in many aspects, the chain will behave as an ideal chain. An important consequence is that the chain size will scale with the square root of the molecular weight instead of the power 0.6 (for D = 3) of a SAW.

3.2.4 A REAL CHAIN IN THE MELT

Now we address the question of the configurations of a chain in its own melt. In first instance, one is tempted to state that this is equivalent to an athermal solvent and the polymer chains will behave as SAWs. In particular, their size will scale with $M^{0.6}$. This turns out to be wrong. Already in 1949, Flory stated that chains in a melt are gaussian and ideal.[9] This notion remained debated until neutron scattering experiments[10,11] definitely showed that indeed, chain sizes in the melt scale with $M^{0.5}$. The reason is that an individual polymer coil expands because of the intra-molecular interactions. In the melt there is no difference between intra and inter molecular interactions: a chain segment does not "know" whether it is interacting with a remote segment of the same molecule or with another molecule. This goes for all chains and hence there is no net swelling force.[6]

The implications of the concept can be illustrated by the following example. We depict the polymer as a very long and thin coiled balloon. If uninflated, the balloon is infinitely thin. If inflated, it occupies a certain volume. A melt is represented by a volume filled with a random mixture of balloons, much like a plate of spaghetti, such that the inflated balloons would densely pack in the volume. Empty balloons have no volume of their own and will thus behave ideal and their coil size will scale with the square root of their length. A dilute athermal solution corresponds with an isolated balloon in some inert medium, say air or water. Also, this balloon will behave ideal when uninflated.

Now, we turn on excluded volume interactions by inflating the balloons. An isolated balloon will expand as a result of this and will behave as a self avoiding walk. On the other hand, a balloon in the "melt" cannot expand because the other balloons, also inflated, will prevent it from doing so. They will just pack more densely as they are being inflated without changing their configuration. In other words, they will still behave as ideal balloons. If one would suddenly remove (or puncture) all balloons except for one, that last remaining balloon will expand (after a bit of shaking) and behave as a SAW. Even if all balloons, except one, would fall apart in many small balloons, the one remaining would expand. To the best of our knowledge, these experiments, though much less expensive then neutron scattering experiments have never been actually performed. See also Chapter 5.

3.3 IDEAL POLYMER MIXTURES

3.3.1 INTRODUCTION

We wish to describe the miscibility of polymeric mixtures. As discussed in Chapter 2, this is equivalent to finding an expression for the Gibbs free energy of mixing ΔG_M. The notion of an ideal mixture has proven to be a cornerstone of the theory of "small molecule" mixtures so it is worthwhile to explore the same notion for polymeric mixtures.

3.3.2 IDEAL POLYMERIC MIXTURES

Consider a mixture of n_1 and n_2 molecules of different type. The Gibbs free energy of mixing of an ideal mixture is entirely entropic with the entropy of mixing ΔS_M given by:

$$\frac{\Delta S_M}{NR} = -[X_1 \ln X_1 + X_2 \ln X_2] \qquad [3.26]$$

where X are mole fractions ($X_1 + X_2 = 1$) and $N = n_1 + n_2$ is the total number of moles of molecules. The main difference between polymers and "normal" molecules is that polymer molecules occupy a volume which is orders of magnitude larger. However, the molecular volume does not show up in Eq. (3.26), so Eq. (3.26) is in principle applicable to polymer mixtures too. However, Eq. (3.26) gives the free energy of mixing per mole of molecules and a unit volume contains orders of magnitude more "small" molecules than polymers. Therefore, the entropy of mixing per unit volume (or unit weight) is orders of magnitude smaller for mixtures of polymers.

In a polymer solution, that is a mixture of polymers and small molecules, there is a huge size difference between the constituents and the derivation that leads to Eq. (3.26) no longer applies. One may also wonder whether the chain character of polymers plays any role.

These issues were addressed, independently by Flory[12,13] and Huggins,[14,15] in the 1940s. An expression for the entropy of mixing of polymers with different degrees of polymerization (number of segments) r_1 and r_2 was derived:

$$\frac{\Delta S_M}{NR} = -\left[\frac{\Phi_1}{r_1} \ln \Phi_1 + \frac{\Phi_2}{r_2} \ln \Phi_2\right] \qquad [3.27]$$

where Φ represents volume fractions and N is the total number of segments (assumed equal in size). Eq. (3.27) is the usual formulation of the Flory-Huggins entropy of mixing. For a mixture of polymers that are equal in size one obtains the ideal solution entropy of mixing, Eq. (3.26). This can be verified by substituting $r_1 = r_2 = r$ in the expressions for Φ and N:

$$\Phi_i = \frac{n_i r_i}{N} \qquad\qquad [3.28]$$

$$N = n_1 r_1 + n_2 r_2$$

The significance of the Flory-Huggins entropy of mixing lies in mixtures of molecules with large size differences. Indeed, the relation was originally derived for polymer solutions ($r_2 = 1$).

Although the Flory-Huggins entropy of mixing is the cornerstone of most theories on polymer thermodynamics, its connotation is not always recognized. Therefore, it is important to consider its origin in some more detail. In the next section, we will reproduce the derivation of Flory, which makes explicit use of the chain character of the polymer. Subsequently, we show that the same equation can be derived without any reference to a chain.

3.3.3 DERIVATION OF THE FLORY-HUGGINS ENTROPY OF MIXING

The calculations are based on a lattice model with N lattice points (coordination number z) on which n_1 linear polymers consisting of r segments (i.e. lattice points) are to be placed with n_2 solvent molecules (1 per lattice point), $N = n_1 r + n_2$. The n_1 polymers are placed on the lattice and the number of ways this can be done is counted. When m polymers have been added to the lattice, mr sites are occupied. Then the (mean field) assumption is made that these mr polymer segments are evenly distributed over the volume so that the probability P^{free} for an arbitrary site to be empty is given by:

$$P^{free} = \frac{N - mr}{N} \qquad\qquad [3.29]$$

The assumption of a uniform segment density is clearly incorrect for polymers. Especially for dilute solutions when the segments are distributed as isolated islands of chains, this approximation fails. The probability of finding an arbitrary site to be empty is correctly given by Eq. (3.29), only the probabilities for nearby sites are highly correlated. Huggins proposed a correction for this correlation which we shall discuss later.

The reasoning proceeds as follows. The number of possible sites for the first segment of the polymer number (m+1) to be placed on the lattice (N−mr). The (average) number of possible sites for the second segment is zP^{free}. The remaining segments all have $(z-1)P^{free}$ possibilities. In this way we find for the total number of ways v_{m+1} to accommodate the (m+1)th polymer:

$$v_{m+1} = (N - mr)z(z - 1)^{r-2} \left(\frac{N - mr}{N} \right)^{r-1}$$ [3.30]

The left hand side of Eq. (3.30) should actually be multiplied by $\frac{1}{2}$ to correct for the fact that each configuration is counted twice: starting from one end and starting from the other end. This does, however, not affect the final result (Eq. (3.34)) and was also not considered in the original derivation.

The most important approximations in Eq. (3.30) are the above mentioned assumption of a uniform segment density and the modelling of the chain configurations by a random walk without immediate self-reversal. The latter is a result of allowing the chain to intersect itself (no long range intra chain interactions, i.e., a random walk) and taking $z-1$ for the number of potential sites for the $i > 1$th segment (no immediate self-reversals). The total number of different ways Ω to fill the lattice with the mixture is given by:

$$\Omega = \frac{1}{n_1!} \prod_{m=0}^{n_1} v_{m+1}$$ [3.31]

The entropy of the mixture $S = R \ln \Omega$ can then be calculated with minor mathematical approximations and is given by:

$$\frac{S}{R} = -\left[n_1 \ln y_1 + n_2 \ln y_2 - n_1(r - 1) \ln \left(\frac{z - 1}{e} \right) \right]$$ [3.32]

where $y_i = n_i/N$ and $z(z-1)^{r-2}$ has been replaced by $(z-1)^{r-1}$ in Eq. (3.30). In order to obtain the entropy of mixing, the entropies of the pure phases should be subtracted. The entropy of the pure solvent can be found by taking $n_1=0$, $N=n_2$ in Eq. (3.31) to be zero. However, the entropy of the polymer phase does not vanish. Substitution of $n_2=0$, $N=n_1r$ yields:

$$\frac{S^{pol}}{R} = -n_1 \left[\ln r + (r - 1) \ln \left(\frac{z - 1}{e} \right) \right]$$ [3.33]

The first term is the combinatorial entropy of the head segments and the second term is the configurational entropy in the polymer chain. The latter is identical to the corresponding term in Eq. (3.32). This is a consequence of the mean field approximations that the configurational freedom of a polymer chain is independent of the environment. This causes the chain character of the polymer and the lattice coordination number to drop out of the entropy of mixing expression that is obtained by subtracting S^{pol} from S in Eq. (3.32):

$$\frac{\Delta S_M}{R} = - [n_1 \ln \Phi_1 + n_2 \ln \Phi_2] \qquad [3.34]$$

where $\Phi_1 = n_1 r/N$ and $\Phi_2 = n_2/N$ are the volume fractions of respectively polymer and solvent. Eq. (3.34) can alternatively be written as:

$$\frac{\Delta S_M}{NR} = - \left[\frac{\Phi_1}{r} \ln \Phi_1 + \Phi_2 \ln \Phi_2 \right] \qquad [3.35]$$

which is the most frequently used formulation for the Flory-Huggins entropy of mixing (cf Eq. (3.27)).

3.3.4 ALTERNATIVE DERIVATIONS

The Flory-Huggins entropy of mixing does not contain parameters that originate from the lattice model. Therefore, one might expect that Eq. (3.35) can also be derived without reference to a lattice. Such a derivation has indeed been given by Hildebrand,[2,16] based on the following expression for the entropy of n_i particles:

$$\frac{S_{n_i}}{R} = n_i \ln [n_i(v_i - v_i^*)] \qquad [3.36]$$

where $v_i = V/n_i$ is the available volume per particle and v_i^* is the actual geometric volume of the particle. Assuming additivity of the v^*'s in the mixture, a fixed ratio v_i/v_i^* and, just as in the above derivation, no volume change on mixing, one obtains Eq. (3.35) again for the entropy of mixing. This derivation does not use a lattice and does not assume any particular shape (e.g., flexible chain) of the particles.

Yet another derivation can be given using a lattice but without reference to a chain. To this end we fill N sites with n_2 fixed objects where each object occupies r sites and count the number of possible configurations. The first object has a choice out of N sites (we only need to define the location of the first segment since the object is fixed). The first segment of the second object has $N-r$ possible sites. However, all the other $r-1$ segments should also fall on an empty site. Each site has a probability of $(N-r)/N$ to be empty. Hence the total number of possibilities v_2 for the second object is given by:

$$v_2 = (N - r) \left(1 - \frac{r}{N} \right)^{r-1} \qquad [3.37]$$

In general, after placing m objects, the number of possibilities v_{m+1} to accommodate object $m+1$ is given by:

$$v_{m+1} = (N - mr)\left(1 - \frac{r}{N}\right)^{r-1} \qquad [3.38]$$

One then proceeds along the same lines as from Eq. (3.31) downward and obtains again the Flory-Huggins entropy of mixing expression.

The above considerations lead to the conclusion that the Flory-Huggins entropy of mixing expression is not specific for chain molecules. It is the combinatorial entropy of the centers of gravity of the components taking into account the different sizes of the constituents. It is essential that the intramolecular configurational freedom of the components does not change in going from the pure phase to the mixture so that no contribution to the mixing entropy is made. In Hildebrand's and the last derivation this assumption is made explicitly while in the Flory-Huggins theory it is a consequence of modelling the polymer configurations by a random walk.

3.3.5 HUGGINS CORRECTION

Huggins proposed a correction for the correlation of the probabilities to find empty sites due to the chain-like character of the polymer. The probability that the site for the first segment of a chain is empty is given by the fraction P^{free} of empty sites. In the construction, followed above, the next segment is chosen in one of the z potential directions and the probability that this site is empty is also taken as P^{free}. This, however, is an underestimation. The correlation between occupied sites in a chain also implies a correlation between empty sites. The fact that the previous site in a polymer chain was apparently empty increases the probability that the next site is also empty.

It can be shown that the probability $P^{free}(S,S-1)$ that site S is empty, given the fact that the neighboring site S–1 is also empty is given by:

$$P^{free}(S,S-1) = \frac{N - mr}{N - 2(r - 1)\dfrac{m}{z}} \qquad [3.39]$$

The resultant contribution ΔS^H of this effect to the expression for the entropy of mixing is rather involved. However, the expression may be developed into powers of 1/z. The zero order term vanishes: for infinite z one retains the mean field Flory-Huggins result. This can already be seen from Eq. (3.39). The first order term yields:

$$\frac{\Delta S^H}{NR} = -\frac{1}{z}\Phi_1\Phi_2\left(1 - \frac{1}{r}\right)^2 \qquad [3.40]$$

The Huggins correction thus provides a negative (unfavorable) contribution to the entropy of mixing which decreases asymptotically to a value of $-1/z$ for infinite molecular weight. Contrary to the Flory-Huggins expression, Eq. (3.39) shows the use of a lattice model through the lattice coordination number z. We stress however, that the topological connectivity of the lattice plays no role. Basically these are "fixed number of sites, fixed number of neighbor" models.

The Huggins correction is maximal for a symmetric mixture, increases with increasing chain length and vanishes for infinite coordination number z since an infinite coordination space has no effect anymore on configurational entropy, because the number of realizations for a chain is also infinite, irrespective of its environment. On the other hand, the smaller the coordination number and hence, the smaller the number of realizations for a chain, the larger the correction.

3.4 REGULAR POLYMER MIXTURES

In the previous sections it is shown that the Flory-Huggins entropy of mixing, Eq. (3.27), is very general, does not require any specific model assumptions, does not even require the polymers to be chains. In the spirit of ideal solution theory, discussed in Section 2.3, we defined an ideal polymer mixture as one that has a free energy of mixing which is entirely entropic, with the entropy of mixing given by:

$$\frac{\Delta S_M}{NR} = -\left[\frac{\Phi_1}{r_1}\ln\Phi_1 + \frac{\Phi_2}{r_2}\ln\Phi_2\right] \qquad [3.41]$$

Introduction of a regular polymer mixture would be the next logical step. Indeed, the concept has been introduced and is the starting point of almost all subsequent theories. One could safely state that the resulting expression, known as the Flory-Huggins free energy of mixing still is the cornerstone of polymer thermodynamic theories.

The heat of mixing of a regular mixture was discussed in Section 2.4 and is given by:

$$\frac{\Delta H_M}{NRT} = \chi X_1 X_2 \qquad [3.42]$$

where X_i are the mole fractions of molecules 1 and 2 which were assumed to be approximately equal in size. The interaction parameter χ is given by (Eq. (2.79)):

$$\chi = \frac{zW}{RT} \qquad [3.43]$$

where W describes the differences in interaction energies of the segments (see Eq. (2.77)). If the molecules are not equal in size, as is obviously the case in polymer solutions, one could either follow the arguments in Section 2.5 or use a lattice model and interpret X_i as the mole fraction of sites (equal in size) occupied by (a segment of) molecule i. The latter approach was followed by Flory and Huggins. Both approaches lead to the same result: that mole fractions X should be replaced by volume fractions Φ. This brings us at the Flory-Hugging's expression for the Gibbs free energy of mixing:

$$\frac{\Delta G_M}{NRT} = -\left[\frac{\Phi_1}{r_1}\ln\Phi_1 + \frac{\Phi_2}{r_2}\ln\Phi_2\right] + \chi\Phi_1\Phi_2 \qquad [3.44]$$

where the interaction parameter χ has an inverse temperature dependence:

$$\chi = \frac{\chi_h}{T} \qquad [3.45]$$

with χ_h a constant with unit Kelvin. Eq. (3.44) defines what one could call a regular polymer mixture. Indeed, for $r_1 = r_2$, mole fractions and volume fractions are identical and Eq. (3.44) is equivalent to the definition, Eq. (2.78), of a regular solution.

In contrast to the expression for the entropy of mixing, the chain-like character of a polymer does play a role here. We took a lattice point as a basic segment and assumed that each segment was free to interact with z neighboring segments (Eq. (3.43)). This would clearly be incorrect, particularly for polymer solutions, if polymers would be dense solid objects of many segments. Let the solvent molecule be the basic segment with z interaction sites. If the polymer would be a dense object with σzr instead of zr interaction sites with $\sigma < 1$, one may again follow the route leading to Eq. (3.43), counting the number of interactions, to arrive at:

$$\frac{\Delta H_M}{N} = zW\frac{\sigma}{1 - (1 - \sigma)\Phi_1}\Phi_1\Phi_2 \qquad [3.46]$$

Eq. (3.46) is more generally valid for mixtures of molecules with different interaction surfaces s_1 and s_2 ($\sigma = s_1/s_2$) as first noted by Staverman.[17] Interaction surfaces may be estimated, using the group contribution scheme of Bondi.[18] Note that N is the total number of segments (lattice sites). It is convenient to eliminate the rather arbitrary notion of a segment by dividing both sides of Eq. (3.44) by the molar volume V_m of a segment and using the relation between the interaction parameter and solubility parameters, Eq. (2.84). One then obtains:

$$\frac{\Delta G_M}{VRT} = -\left[\frac{\Phi_1}{V_1} \ln \Phi_1 + \frac{\Phi_2}{V_2} \ln \Phi_2\right] + (\delta_1 - \delta_2)^2 \Phi_1 \Phi_2 \qquad [3.47]$$

where $V_i = r_i V_m$ is the molar volume of molecule i and $V = N V_m$ is the total volume. The solubility parameters can be estimated with group contribution methods. Eq. (3.47) thus enables calculation of the thermodynamic phase behavior of two molecules on the basis of their chemical structures and molar volumes.

3.4.1 GENERALIZED REGULAR POLYMER MIXTURES

The Flory-Huggins equation for the free energy of mixing, Eq. (3.44) with Eq. (3.45) for χ soon proved to be inadequate for a quantitative or even qualitative description of polymeric phase equilibria. The most obvious step to take to give the equation more flexibility is to treat χT as a free energy rather than an enthalpy parameter. This implies that χ may have an arbitrary temperature dependence. As a more general extension of the meaning of the interaction parameter, we may write:

$$\chi = \chi_s + \frac{\chi_h}{T} \qquad [3.48]$$

The Huggins correction, Eq. (3.40), can then be incorporated as an entropic correction χ_s to χ:

$$\chi_s = \frac{1}{z}\left(1 - \frac{1}{r}\right)^2 \qquad [3.49]$$

Since the composition dependence of the free energy of mixing is not altered by this generalization, the thermodynamic analysis is also basically the same with modified temperature scales. We could call such models generalized regular polymer mixtures.

3.5 PHASE DIAGRAMS

We are now in the position to calculate phase diagrams of polymer blends and solutions. One may calculate the chemical potential $\Delta\mu_1$ of polymer 1 in a mixture of composition Φ_1, Φ_2 relative to its pure phase according to Eq. (2.10):

$$\Delta\mu_1 = \left(\frac{\partial \Delta G_M}{\partial n_1}\right)_{n_2} \qquad [3.50]$$

Using Eqs (2.27-2.28) and Eq. (3.44) one then obtains:

$$\frac{\Delta\mu_1}{RT} = \ln\Phi_1 + \left(1 - \frac{r_1}{r_2}\right)\Phi_2 + r_1\chi\Phi_2^2 \qquad [3.51]$$

and:

$$\frac{\Delta\mu_2}{RT} = \ln \Phi_2 + \left(1 - \frac{r_2}{r_1}\right)\Phi_1 + r_2\chi\Phi_1^2 \qquad [3.52]$$

If two phases, denoted I and II, with different compositions (Φ_1^I, Φ_2^I) and (Φ_1^{II}, Φ_2^{II}) are in thermodynamic equilibrium then the chemical potentials of each component should be equal in both phases:

$$\Delta\mu_1^I = \Delta\mu_1^{II} \qquad [3.53]$$

$$\Delta\mu_2^I = \Delta\mu_2^{II}$$

With Eqs. (3.51) and (3.52) one obtains:

$$\ln \Phi_1^I + \left(1 - \frac{r_1}{r_2}\right)\Phi_2^I + r_1\chi(\Phi_2^I)^2 = \ln \Phi_1^{II} + \left(1 - \frac{r_1}{r_2}\right)\Phi_2^{II} + r_1\chi(\Phi_2^{II})^2$$

$$[3.54]$$

$$\ln \Phi_2^I + \left(1 - \frac{r_2}{r_1}\right)\Phi_1^I + r_2\chi(\Phi_1^I)^2 = \ln \Phi_2^{II} + \left(1 - \frac{r_2}{r_1}\right)\Phi_1^{II} + r_2\chi(\Phi_1^{II})^2$$

Eq. (3.54) constitutes a set of coupled equations that must be solved numerically because of the combination of logarithmic and non-logarithmic terms. A suitable approach to calculate the binodal curve is the eliminate χ from Eqs. (3.54). This yields one equation with two independent variables Φ_1^I and Φ_1^{II} ($\Phi_2 = 1 - \Phi_1$). Then take Φ_1^I fixed and solve numerically, e.g., by using a Newton-Raphson scheme, for Φ_1^{II}. Subsequently, χ can be found by substituting the values for Φ_1^I (as taken) and Φ_1^{II} (as calculated) in either one of the two equations in Eq. (3.54). From the temperature dependence of χ, e.g., Eq. (3.48) but any temperature function may be used, one obtains the binodal (cloud point) temperature of a mixture with composition Φ_1^I and the composition Φ_1^{II} of the incipient phase. The spinodal, i.e., the boundary between thermodynamically metastable and unstable compositions is more easy to calculate as it is given by the solution of (see Eq. (2.44):

$$\frac{\partial^2 \Delta G_M}{\partial \Phi_1^2} = 0 \qquad [3.55]$$

Substitution of Eq. (3.44) yields:

$$\frac{1}{r_1\Phi_1} + \frac{1}{r_2\Phi_2} = 2\chi \qquad [3.56]$$

In addition, the critical point of the mixture should obey the condition:

$$\frac{\partial^3 \Delta G_M}{\partial \Phi_1^3} = 0 \tag{3.57}$$

which yields an expression for the critical concentration $(\Phi_1)_C$:

$$(\Phi_1)_C = \frac{1}{1 + \left(\dfrac{r_1}{r_2}\right)^{\!\frac{1}{2}}} \tag{3.58}$$

Equation (3.58) represents one of the more eminent successes of the FH model, namely the correct description of the shape of the phase diagram which is skewed towards the solvent-rich side ($\Phi_1 = 0$) for polymer solutions ($r_1 \gg r_2$). The physical implication is that a phase separated polymer solution will generally consist of almost pure solvent in equilibrium with a solvent swollen polymer phase.

Since the critical point is located on the spinodal we may substitute Eq. (3.58) into Eq. (3.56) to obtain a critical value χ_C of the interaction parameter:

$$\chi_C = \frac{1}{2}\left(\frac{1}{r_1^{\frac{1}{2}}} + \frac{1}{r_2^{\frac{1}{2}}}\right)^2 \tag{3.59}$$

If $\chi < \chi_C$ the mixture is always miscible, if $\chi > \chi_C$ the mixture may phase separate into two coexisting phases as defined by the binodal.

Let us investigate the influence of chain lengths r_1, r_2 on phase stability in somewhat more detail. We will assume that the interaction parameter χ is independent of chain length and has a simple $1/T$ temperature dependence, Eq. (3.45) with $\chi_h > 0$; for $\chi_h \leq 0$, the constituents are always miscible. Binodals and spinodals of low molecular weight mixtures ($r_1 = r_2 = 1$), polymer solutions (r_1 large, $r_2 = 1$) and polymer mixtures (r_1, r_2 large) are shown in Figure 3.6. Note that all phase diagrams describe phase separation upon cooling, upper critical demixing. The critical points are located at the maxima of the binodals. These are the upper critical solution temperatures (UCST). Low molecular weight mixtures ($r_1 = r_2 = 1$) have a relatively large temperature region of complete miscibility; UCST = $\chi_h/2$. Polymeric mixtures (r_1, $r_2 \to \infty$) show hardly any miscibility in a relevant temperature region; UCST $\to \infty$. Polymer solutions ($r_1 \to \infty$, $r_2 = 1$) are in-between cases UCST = $2\chi_h$, and the larger the polymer chain length, the more the phase diagram is skewed towards the solvent-rich phase.

For polymer solutions, in the limit $r_1 \to \infty$ we have $(\Phi_1)_C \to 0$ and $\chi_C \to 1/2$. This special value for χ marks the point in the phase diagram where a polymer solution with infinite polymer molecular weight will

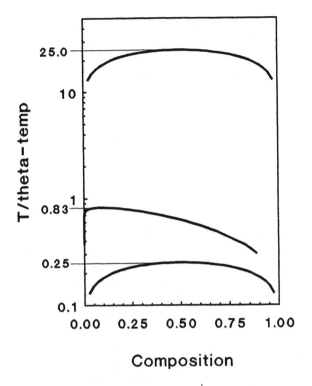

Composition

Figure 3.6. Typical UCST phase diagrams of liquid mixtures ($r_1 = r_2 = 1$), polymer solutions ($r_1 = 100$, $r_2 = 1$), and polymer blends ($r_1 = r_2 = 100$).

start to phase separate. This is true, irrespective of the particular temperature dependence of χ. For a decreasing function $\chi(T)$ as in $\chi = \chi_h/T$, $\chi = 1/2$ marks an UCST and for an increasing function $\chi(T)$, $\chi = 1/2$ indicates an LCST. For $\chi > 1/2$ dilute solutions will demix into a swollen polymer and a solution with (infinitely) low polymer concentration.

The case $\chi = 1/2$ is also special because a dilute polymer solution behaves as an ideal solution under this condition. This can be seen as follows. Consider the chemical potential $\Delta\mu_2$ of the solvent ($r_2 = 1$) in a polymer solution, Eq. (3.52):

$$\frac{\Delta\mu_2}{RT} = \ln\Phi_2 + \left(1 - \frac{1}{r_1}\right)\Phi_1 + \chi\Phi_1^2 \qquad [3.60]$$

For $\chi = 0$ and $r_1 = 1$ one obtains $\Delta\mu_2/RT = \ln\Phi_2 = \ln X_2$ which is the result for an ideal mixture, see Eq. (2.63). For $r_1 > 1$ one finds from numerical evaluation that the chemical potential is less then ideal. Osmotic and vapor pressures will also be less than those of an ideal mixture. Using Eq. (3.28), with $r_2 = 1$, we may write the chemical potential in Eq. (3.60) as the sum of the ideal term and an excess term:

$$\frac{\Delta\mu_2}{RT} = \ln X_2 + \ln\left[1 - \left(1 - \frac{1}{r_1}\right)\Phi_1\right] + \left(1 - \frac{1}{r_1}\right)\Phi_1 + \chi\Phi_1^2 \qquad [3.61]$$

For small Φ_1 the logarithm may be expanded. The first order term vanishes against the third term in Eq. (3.61) and up to second order in Φ_1 we obtain:

$$\frac{\Delta\mu_2}{RT} = \ln X_2 + \left[\chi - \frac{1}{2}\left(1 - \frac{1}{r_1}\right)^2\right]\Phi_1^2 \qquad [3.62]$$

Eq. (3.62) shows that a high molecular weight ($r_1 \to \infty$) polymer solution behaves as an ideal solution at infinite dilution, as it should, and also at finite concentrations under the condition $\chi = 1/2$. The special temperature where $\chi = 1/2$ is called the theta (Θ) temperature.

Note that the considerations above follow from the Flory-Huggins free energy of mixing which, as we know, does not need polymers to be chains. A single polymer chain of infinite length can itself be regarded as a dilute solution of segments. At the theta point, defined by $\chi = 1/2$, this solution is ideal. This implies that solvent-solvent, solvent-segment and segment-segment interactions are identical (the energetic interactions are not as $\chi > 0$, but at the particular temperature defined by $\chi = 1/2$, these interactions are compensated by the excess entropic interactions). This means that the polymer segments are effectively only constrained by their connectivity and the chain thus behaves as an ideal chain, see also Chapter 5.

Another way of putting this is to write the virial expansion of the osmotic pressure Π:

$$\Pi = RTc\left(\frac{1}{M_n} + A_2 c + \ldots\right) \qquad [3.63]$$

With Eq. (3.60) for the chemical potential we obtain:

$$\Pi = RTc\left[\frac{1}{M_n} + \left(\frac{v_{pol}^2}{V_{sol}}\right)\left(\frac{1}{2} - \chi\right)c + \left(\frac{v_{pol}^3}{6V_{sol}}\right) + \ldots\right] \qquad [3.64]$$

where c is the concentration of polymer, v_{pol} is the specific volume of the polymer and V_{sol} is the molar volume of the solvent. Comparison of Eq.(3.63) and Eq.(3.64) yields:

$$A_2 = \left(\frac{v_{pol}^2}{V_{sol}}\right)\left(\frac{1}{2} - \chi\right) \qquad [3.65]$$

which shows that at the theta point the second virial coefficient A_2 is zero.

In the previous sections we saw that $\chi = 1/2$ also defined the critical solution temperature of an infinite molecular weight polymer solution. For $\chi > 1/2$, $T < \Theta$ temperature for a USCT. The polymer solution will demix into a swollen polymer and almost pure solvent. Again, this may be translated to the behavior of one infinite molecular weight polymer coil. Such a coil will shrink to a collapsed state for $\chi > 1/2$, driving out solvent, see Chapter 6.

3.6 THE EFFECT OF MOLECULAR WEIGHT DISTRIBUTIONS

3.6.1 INTRODUCTION

The vast majority of practical polymers are a mixture of polymers of different molecular weights. In the foregoing we were assuming infinitely small molecular weight distributions (MWDs). In this section, we discuss the effects of a finite MWD on the phase equilibria. Polymers with a significantly finite MWD are often also referred to as polydisperse polymers. Strictly speaking, each chain length (molecular weight) represents one component. A high molecular weight polydisperse polymer thus consists of a very large, virtually infinite, number of components. We designate the volume fraction of component i by the symbol ψ_i. The generalization of the Flory-Huggins free energy of mixing for a mixture of N such components is:

$$\frac{\Delta G_M}{RT} = \sum_{i=1}^{N} \frac{\psi_i}{r_i} \ln \psi_i + \Gamma(\psi_1, \psi_2, ..., \psi_N, T) \qquad [3.66]$$

Equation (3.41) is regained for N=2 and $\Gamma(\psi_1, \psi_2, T) = \chi \psi_1 \psi_2$. In thermodynamic equilibrium, when two phases I and II are present, the following equations must be satisfied:

- equality of chemical potentials:

$$\mu_j^I = \mu_j^{II} \quad (j = 1,..,N) \qquad [3.67]$$

- compositions in both phases must add up to the total composition ψ_j^F (F denotes "feed"):

$$\nu \psi_j^I + (1 - \nu)\psi_j^{II} = \psi_j^F \quad (j = 1,...,N) \qquad [3.68]$$

where ν is the volume fraction occupied by phase I.

Finally, we must demand that the volume fractions ψ_j^I and the volume fractions ψ_j^{II} both add up to one. Assuming that the feed composition ψ^F is normalized properly we see with Eq. (3.65) that it suffices to demand that the sums of the volume fractions in both phases be equal:

$$\sum_{j=1}^{N} (\psi_j^I - \psi_j^{II}) = 0 \qquad\qquad [3.69]$$

Eqs (3.67÷3.69) constitute a set of 2N + 1 equations with 2N + 2 unknowns (2N volume fractions, T and v). In order to define a unique thermodynamic state, one must therefore add one extra equation, usually the specification of one of the unknown variables. For isothermal "flash" calculations one specifies the temperature T. If this temperature is in the two phase region for a composition ψ^F, a solution exists for $0 < v < 1$. At the cloud point (or binodal) one of the phases, say II, is present in an infinitesimal amount. In this case one specifies $v = 1$. The composition of phase II is known as the shadow point.

The expression for the spinodal of an N component system has been formulated by Gibbs[17] and can be written as:

$$\left| \frac{\partial^2 \Delta G_M}{\partial \psi_i \partial \psi_j} \right| = 0 \qquad\qquad [3.70]$$

which states that the determinant of the (N–1)x(N–1) matrix with elements i,j as given by Eq. (3.70) is zero. One of the N volume fractions, say ψ_1 must be treated as dependent variable: $\psi_1 = 1 - \Sigma_{j=2..N} \psi_j$. Calculation of spinodals thus involves the evaluation of N–1xN–1 determinants.

For most practical systems, N will be extremely large and exact direct solution of the phase equilibria equations becomes unfeasible. However, for a certain broad class of functions the equations can be reduced to a much smaller set of equations which can readily be solved.[18,19]

The method works for all functions that can be written as:

$$\Gamma(\psi_1, \psi_2,..,\psi_N, T) = \Gamma(Q_1, Q_2,..., Q_K, T) \qquad\qquad [3.71]$$

where the set of K functions Q depend linearly on the ψ_i:

$$Q_s = \sum_i q_{si}\psi_i \qquad\qquad [3.72]$$

In order to apply this method to polymers with MWDs we introduce the concept of a polymer family. A polymer family roughly means a group of polymers that consist of the same monomers, such that the interaction function Γ in Eq. (3.63) is a function of K family volume fractions Φ_i only:

$$\Gamma(\psi_1, \psi_2,..., \psi_N, T) = \Gamma(\Phi_1, \Phi_2,.., \Phi_K, T) \qquad [3.73]$$

with:

$$\Phi_i = \sum_{j=\Omega_i} \psi_j \qquad [3.74]$$

where Ω_i denotes the collection of labels {j}, for which component j is a member of family i. As an example, consider a blend of PVC and PMMA, both polymers having a MWD. Such a blend may be considered to consist of two chemically different families (PVC and PMMA chains) with a segmental Flory-Huggins interaction parameter χ. We then have:

$$\Gamma(\psi_1,.., \psi_N, T) = \chi\Phi_1\Phi_2 \qquad [3.75]$$

where Φ_1 and Φ_2 are the overall volume fractions of respectively PVC and PMMA.

3.6.2 SPINODAL AND CRITICAL POINT

From the more general mathematical treatment,[19] the following results may be derived for a blend of two polydisperse polymer families. The equation for the spinodal is:

$$\frac{1}{r_{w1}\Phi_1} + \frac{1}{r_{w2}\Phi_2} = 2\chi \qquad [3.76]$$

where r_{wi} is the weight averaged molecular weight of the molecules in family i. The location of the spinodal thus only depends on the weight averaged molecular weights of the components. The form of the spinodal equation is identical to Eq. (3.56) for a strictly binary mixture with chain lengths replaced by weight averaged chain lengths. The expression for the critical point is given by:

$$\frac{r_{z1}}{r_{w1}^2\Phi_1^2} = \frac{r_{z2}}{r_{w2}^2\Phi_2^2} \qquad [3.77]$$

This can be written as:

$$(\Phi_1)_c = \frac{1}{1 + \left(\dfrac{r_{w1}}{r_{w2}}\right)^{1/2}\left(\dfrac{Q_{zw2}}{Q_{zw1}}\right)^{1/2}} \qquad [3.78]$$

where Q_{zwi} is the heterogeneity index r_{zi}/r_{wi}. This expression reduces to Eq. (3.58) for the monodisperse case ($Q_{zwi} = 1$). A polydisperse polymer mixture thus has a spinodal determined by the weight averaged molecular weights only. The top of the spinodal is given by

Eq. (3.58) with chain lengths replaced by weight averages. Eq. (3.78) shows that the critical point is shifted in the direction of the composition axis corresponding to the component with the highest Q_{zw}. Similar expressions can be derived for mixtures with more than two "families". For three families:

$$\Gamma = \chi_{12}\Phi_1\Phi_2 + \chi_{13}\Phi_1\Phi_3 + \chi_{23}\Phi_2\Phi_3 \qquad [3.79]$$

the spinodal equation is given by:

$$\frac{r_{w1}\Phi_1}{1 + (\chi_{23} - \chi_{12} - \chi_{13})r_{w1}\Phi_1} + \frac{r_{w2}\Phi_2}{1 + (\chi_{13} - \chi_{12} - \chi_{23})r_{w2}\Phi_2} +$$

$$+ \frac{r_{w3}\,\Phi_3}{1 + (\chi_{12} - \chi_{13} - \chi_{23})r_{w3}\,\Phi_3} = 0 \qquad [3.80]$$

and the critical point follows from eq. (3.78) and:

$$\frac{r_{w1}r_{z1}\Phi_1}{1 + (\chi_{23} - \chi_{12} - \chi_{13})r_{w1}\Phi_1} + \frac{r_{w2}r_{z2}\Phi_2}{1 + (\chi_{13} - \chi_{12} - \chi_{23})r_{w2}\Phi_2} +$$

$$+ \frac{r_{w3}r_{z3}\,\Phi_3}{1 + (\chi_{12} - \chi_{13} - \chi_{23})r_{w3}\,\Phi_3} = 0 \qquad [3.81]$$

Again, the spinodal only depends on the weight averaged molecular weights and the critical point(s) on weight and z-averages.

3.6.2 BINODAL AND FLASH CALCULATIONS

With a free energy of mixing, given by Eq. (3.66), the chemical potential of component j (with volume fraction relative to the pure phase $\Delta\mu_j$ is given by (see Eq. (2.26)):

$$\frac{\Delta\mu_j}{RT} = 1 + \ln\psi_j + r_j\left[\Gamma + \frac{\partial\Gamma}{\partial\Psi_j} - \sum_{k=1}^{N}\frac{\Psi_k}{r_k} - \sum_{k=1}^{N}\psi_k\frac{\partial\Gamma}{\partial\psi_k}\right] \qquad [3.82]$$

We now introduce the distribution factor k_j, defined as the ratio of the component volume fractions between the two phases:

$$k_j = \frac{\psi_j^{II}}{\psi_j^{I}} \qquad [3.83]$$

With the equilibrium condition Eq. (3.67), Eq. (3.82) for the chemical potential, and Eq. (3.73) for Γ, one obtains:

$$k_j = e^{r_j(h_i^I - h_i^{II}) + r_j(h_0^I - h_0^{II})} \qquad [3.84]$$

with:

$$h_0 = \Gamma - \sum_{i=1}^{K} \Phi_i \frac{\partial \Gamma}{\partial \Phi_i} - \Phi_{K+1} \qquad [3.85]$$

$$h_i = \frac{\partial \Gamma}{\partial \Phi_i}$$

and:

$$\Phi_{K+1} = \sum_{j=1}^{N} \frac{\psi_j}{r_j} \qquad [3.86]$$

With the material balance condition, Eq. (3.68), we can now express the compositions of the two coexisting phases as:

$$\psi_j^I = \frac{\psi_j^F}{\nu + (1 - \nu)k_j} \qquad [3.87]$$

$$\psi_j^{II} = \frac{k_j \psi_j^F}{\nu + (1 - \nu)k_j}$$

It can be shown that the set of 2N+1 equations Eq. (3.67÷3.69) can be reduced to the set of K+2 equations:

$$\sum_{j \in Q_i} \frac{\psi_j^F (k_j - \kappa_i)}{\nu + (1 - \nu)k_j} = 0 \quad i = 0,...,K+1 \qquad [3.88]$$

where $\kappa_0 \equiv 1$ and the other κ_i are the family distribution factors:

$$\kappa_i = \frac{\Phi_i^{II}}{\Phi_i^I} \qquad [3.89]$$

Note that Φ_{K+1} is not a family volume fraction but a formal parameter in the theory. We thus need to solve only K+2 equations (3.88) in K+3 unknowns (ν, T and Φ_i, i=1...K+1) instead of 2N+1 equations without any mathematical approximation. The simplification comes entirely from the mathematical construction of the problem. Koningsveld[20] showed for a particular choice of Γ that such a reduction could be achieved. The theory shows that this can be done for a much more general class of interaction functions. The analysis also holds for continuous distributions (continuous thermodynamics) where integrals replace summations.

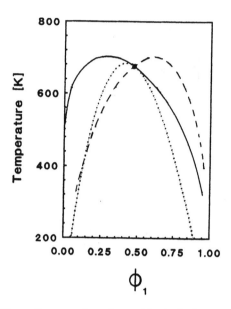

Figure 3.7. Phase diagram of a polymer blend of polymer 1 with Shultz-Flory MWD and polymer 2 with a narrow MWD.

3.6.4 EXAMPLE

As an example we shall discuss the phase diagram of a polymer blend of two polymers with a Shulz-Flory[2] chain length distribution:

$$\psi(r) = r(1-p)^2 p^{r-1} \qquad [3.90]$$

Such a distribution is obtained for "ideal" condensation polymerization.[19] The parameter p is the fraction of bonds that is not at the end of a chain. The average chain lengths are given by:

$$r_n = \frac{1}{1-p} \quad r_w = \frac{1+p}{1-p} \quad r_z = \frac{1+4p+p^2}{1-p^2} \qquad [3.91]$$

For polymer 1, we take p = 0.99. This yields: $r_{n1} = 100$, $r_{w1} = 200$ and $r_{z1} = 300$.

For the calculations we took 100 components with chain lengths r = 1,11,21,...991. As a result the actual cut off and lumped distribution had $r_{n1} = 100$, $r_{w1} = 200$ and $r_{z1} = 300$, which is very close to the theoretical values. For polymer 2 we take r = 100. Figure 3.7 shows the corresponding phase diagram (for $\chi = 10/T$). The full drawn line is the binodal or cloud point curve. The roughly dashed line is the corresponding shadow curve and the finely dashed line is the spinodal. All curves intersect at the critical point, indicated by a dot. One should

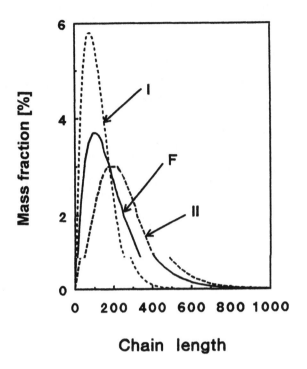

Chain length

Figure 3.8. Chain length distribution in the feed and in both coexisting phases I and II.

always take the shadow point on the curve at the other side of the critical point.

Figure 3.8 shows the result of a flash calculation at $T = 500$ K and $\Phi_1^F = 0.2$. Phase I occupies 87.2% of the total phase volume and contains most of the lower molecular weight material. Phase II contains relatively much of the high molecular weight tail of the distribution. However, the two different fractions do not differ spectacularly: $r_{w1}^I = 124$ and $r_{w1}^{II} = 260$.

REFERENCES

1. R. J. Young, *Introduction to Polymers*, Chapmann and Hall, London (1986).
2. P. J. Flory, *Principles of Polymer Chemistry*, Cornell University Press, Ithaca (NY) (1971).
3. Ref. mbt LCPs
4. P. J. Flory, *Statistics of Chain Molecules*, Interscience Publishers, New York (1969).
5. L. R. G. Trelour, *The Physics of Rubber Elasticity*, Clarendon Press, Oxford (1958).
6. P. G. de Gennes, *Scaling Concepts in Polymer Physics*, Cornell University Press, Ithaca (NY) (1985).

7. D. S. McKenzie, *Polymers and Scaling, Physics Reports*, **27**, 35 (1976).
8. C. J. Adkins, *Equilibrium Thermodynamics*, Cambridge University Press, Cambridge (UK) (1985).
9. P. J. Flory, *J. Chem. Phys.*, **17**, 303 (1949).
10. *Proceedings of the Symposium on the Amorphous State, J. Macromol. Sci.*, **B12** (1976).
11. J. S. Higgins and R. S. Stein, *J. Appl. Crystallogr.*, **11**, 346 (1978).
12. P. J. Flory, *J. Chem. Phys.*, **9**, 660 (1941).
13. P. J. Flory, *J. Chem. Phys.*, **10**, 51 (1942).
14. M. L. Huggins, *J. Chem. Phys.*, **9**, 440 (1941).
15. M. L. Huggins, *Ann. N.Y. Acad. Sci.*, **43**, 1 (1942).
16. J. H. Hildebrand, *J. Chem. Phys.*, **15**, 225 (1947).
17. A. J. Staverman, *Rec. Trav. Chim.*, **56**, 885 (1937).
18. A. Bondi, *J. Phys. Chem.*, **68**, 441 (1964)
19. J. W. Gibbs, *The Collected works of J. Williard Gibbs, Vol. 1: Thermodynamics*, Yale University Press (1948).
20. E. M. Hendriks, *private communication*.
21. R. Koningsveld, *On liquid-liquid phase relationships and fractionation in multicomponent polymer solutions*, PhD thesis, Leiden, 1967.

Chapter 4

POLYMER
THERMODYNAMIC MODELS

4.1 INTRODUCTION

Nowadays, some fifty years after the first papers of Flory and Huggins, a sizable amount of thermodynamic models that describe the phase behavior of more or less complicated polymeric mixtures, has been developed. The more traditional models can be categorized into two classes, namely lattice models and Equation-of-State models. The Flory-Huggins model is the oldest lattice model. In its original form, it is only capable of predicting phase separation phenomena upon cooling (UCST demixing). Many polymeric mixtures as well as polymer solutions show phase separation phenomena upon heating (LCST demixing) as well. It will be discussed in this chapter that such demixing is caused by the compressible nature of polymeric mixtures, by specific energetic interactions between unlike components (such as hydrogen bonds), or by combinations of both.

In Equation-of-State (EoS) models, compressibility effects are modeled by introduction of the "cell" concept: a cell is a lattice site of which the volume depends on relevant model parameters. Such volume represents the free volume available for each site, i.e., polymer segment. The cell concept was successfully introduced first by Prigogine, and was further developed by Flory and several coworkers. In modified versions of EoS models, empty sites on the lattice were introduced. Such models include concepts originally developed by Sanchez and Lacombe. In the so called "Holes and Huggins" model, recently developed by Nies and coworkers, both empty lattice sites and variable cell volume are allowed for.

LCST phase separation phenomena can also be brought about by specific interactions between unlike polymer (or, polymer - solvent) segments. A robust lattice-type description of such effects was developed by Ten Brinke. In this context, Coleman and Painter came up with some practical guidelines to predict (im)miscibility. A generalized model in which both compressibility effects and specific interactions are taken into account was developed by Sanchez and Balasz.

Two important new developments in polymer thermodynamics, that have more recently come into play thanks to interest from theoretical physics – side on the one hand, and ever increasing computer power on the other hand, will be briefly discussed. These are the lattice cluster theory of Freed, which provides an exact lattice solution of the combinatorial entropy of mixing problem, and the novel off-lattice approach developed by Curro and Schweizer.

It is the aim of this chapter not to arrive at a complete overview of all models currently available, but rather to discuss the main features of the most commonly used models, without going extensively into mathematical detail. Emphasis will be on the basic physics employed and on some characteristics behind model parameters encountered.

4.2 FLORY-HUGGINS MODELS

As discussed in the previous chapter, the most elementary Flory-Huggins (FH) model[1-4] for the free energy of mixing of two polymeric species with chain lengths r_1 and r_2 is:

$$\frac{\Delta G_M}{N_A kT} = \frac{\Delta G_M}{RT} = \frac{\Phi_1}{r_1} \ln \Phi_1 + \frac{\Phi_2}{r_2} \ln \Phi_2 + \chi \Phi_1 \Phi_2 \qquad [4.1]$$

Note it is customary, as in Eq. (4.1), to express a free energy of mixing per mole of molecules. The Flory-Huggins interaction parameter χ originates form dispersive interactions exclusively and is defined as:

$$\chi = \frac{V_L}{RT}(\delta_1 - \delta_2)^2 \qquad [4.2]$$

where V_L is the lattice site volume and δ_i are the solubility parameters of the respective constituents.

The basic characteristics of the FH model, in particular the influence of chain lengths r_i on the phase behavior, were discussed in the previous chapter. We will here investigate some practical aspects of the FH model in more detail. We start with a simple, symmetric polymer mixture ($r_1 = r_2$). From Eq. (3.59) we obtain:

$$\chi_c = \frac{2}{r} \qquad [4.3]$$

It follows from Eq. (4.2) that the maximum limit in solubility parameter difference in order to have complete miscibility at a temperature T is given by:

$$\Delta\delta = \left(\frac{2R}{V_L}\right)^{1/2}\left(\frac{T}{r}\right)^{1/2} = K\left(\frac{T}{r}\right)^{1/2} \qquad [4.4]$$

With a typical value for V_L of 100 cm^3/mol, the value for the constant K amounts to 0.2. Hence, to have a miscible two component polymeric melt of chain lengths 400 (number average molecular weight 40,000) above 400K (not an unrealistic processing temperature), the solubility parameter difference should not exceed 0.2 √cal/cc. This difference is approximately equal to the uncertainty limit of calculating pure component solubility parameters. The practical implication is that the solubility parameters of the two components should be matched.

This example clarifies why so few polymeric mixtures are miscible: the molecular weights are too large and the chemical structure of the constituents too distinct to keep the interaction parameter χ sufficiently small. Sometimes, simple polymeric mixtures of sufficiently low molecular weight do reveal an UCST. For example, an UCST of 450K is reported for a symmetric polyisoprene-polystyrene (PIP/PS) blend with M_n = 2700.[5] In order to predict the phase behavior, one is forced to make a choice for the basic lattice segment molecular weight, as well as for the value for V_L. Unfortunately, volumes and weights of the repeat units of PIP and PS have a sizable difference (76 and 120 gram/mol, respectively), which does not fit into the lattice model philosophy. If we take the smallest repeat unit (PIP) as the basic segment, we arrive at a solubility parameter difference of 0.7 √cal/cc. From Small's group contribution scheme we indeed obtain δ_{PIP} = 8.2 and δ_{PS} = 9.1.

Polymer solutions are somewhat more complicated. One of the most investigated polymer solutions is polystyrene (PS) in cyclohexane (CH) with an experimentally determined Θ-temperature of 303K.[6-8] Using a value for V_L equal to the molar volume of CH, which is 108 cc/mol, and a value for the solubility parameter of CH of 8.2 √cal/cc, a Θ-temperature (at χ = 1/2) of only 80K is predicted using Eq. (4.2).

Θ-temperatures in polymer solutions are always largely underestimated when the FH model is used. It implies that χ is predicted too low. For a relevant description of the phase behavior of polymer solutions at least the Huggins correction should be included. Including this correction, Eq. (3.49), into the free energy of mixing expression Eq. (4.1), according to the convention ΔG = –TΔS + ΔH, we obtain:

$$\frac{\Delta G_M}{RT} = \frac{\Phi_1}{r_1} \ln \Phi_1 + \frac{\Phi_2}{r_2} \ln \Phi_2 + \left[\frac{1}{z}\left(1 - \frac{1}{r}\right)^2 + \chi \right] \Phi_1 \Phi_2 \qquad [4.5]$$

which can be rewritten as:

$$\frac{\Delta G_M}{RT} = \frac{\Phi_1}{r_1} \ln \Phi_1 + \frac{\Phi_2}{r_2} \ln \Phi_2 + \left(\chi_s + \frac{\chi_h}{T} \right) \Phi_1 \Phi_2 \qquad [4.6]$$

and where

$$\chi_s = \frac{1}{z}\left(1 - \frac{1}{r}\right)^2 \qquad\qquad [4.7]$$

$$\chi_h = \frac{V_L}{R}(\delta_1 - \delta_2)^2$$

Hence, a phenomenological repair of the underestimation of interaction parameters is obtained by assuming that the interaction parameter χ can be written as the sum of an entropic contribution χ_s and an enthalpic contribution χ_h/T. The entropic contribution bears in principle relevance to the Huggins correction, but is in practice often taken as an adjustable parameter.

Besides the fact that Θ-temperatures are predicted incorrectly by the FH model, the predicted critical concentration and coexisting phase concentrations are usually somewhat in error as well. A possible way to improve upon this aspect in an empirical way is to make the interaction parameter χ a function of concentration Φ, as well. Such concentration dependence is often used by Kongingsveld and Kleintjes (KK).[9-11] A rationale often used behind such concentration dependence is that in general lattice coordination numbers of both polymeric species are not equal: $z_1 \neq z_2$. It is assumed that the lattice coordination number reflects the number of contacts that polymeric segments can make with neighboring segments. In this context, the lattice coordination number is proportional to the molecular surface area σ: $z_1/z_2 = \sigma_1/\sigma_2$. Defining the non-ideality factor γ as $1 - \sigma_2/\sigma_1$, the interaction parameter χ can be written as:

$$\chi = \chi_s + \frac{\beta}{1 - \gamma\Phi_2} \qquad\qquad [4.8]$$

where the term β is given an appropriate temperature dependence to predict UCST demixing:

$$\beta = \beta_0 + \frac{\beta_1}{T} \qquad\qquad [4.9]$$

It was shown that such introduction of more empirical parameters leads to a better matching of model and experiment.[10] It is also claimed that the factor γ (difference in segmental surface areas) bears relevance to molecular surface areas as predicted independently by group contribution schemes, such as that of Bondi.[12]

Summarizing, the FH model is capable of predicting (im)miscibility of simple polymeric mixtures. More precisely, it is capable of predicting basic features of upper critical demixing in polymer blends and polymer solutions. The FH model, as presented in its form described above, cannot predict lower critical demixing. How the model should be modified to include this phenomenon as well, will be the subject of the next paragraphs.

Finally, as discussed in detail in Chapter 3.6, polymer mixtures and polymer solutions cannot be in general treated as true two component systems. In practice, a polymer sample will always consist of a mixture of different chain lengths. Assuming that energetic interactions are independent of chain length (no end-group effects), this affects the combinatorial entropy of mixing only. The most important effects of polydispersity on phase behavior can be briefly summarized as:[13]

- The location of the binodal depends on the full molecular weight distribution (M_n, M_w, M_z) of the constituents.
- The location of the spinodal depends on M_w exclusively.
- The location of the critical point depends on both M_z and M_w, but not on M_n.

The consequence of polydispersity is that a critical point no longer coincides with the top (or bottom) of the phase diagram, but will in general be shifted towards higher polymer concentrations. The effect of polydispersity can in general be neglected when well defined model polymers, typically with $M_z/M_w < 1.1$, are used.

4.3 EQUATION-OF-STATE MODELS

4.3.1 KEY CONCEPTS

Up to this point the problem of mixing two polymeric species was treated as a combinatorial problem concerned with the positions of segments on a fixed lattice, assuming a certain simple form for the interactions between them. We did not consider yet the fact that in reality polymer segments are not fixed to one discrete position, but have some freedom to move around. Polymer segments (actually every molecular component in a liquid) have some "free volume" to wander around their equilibrium positions without being hindered by neighboring segments.

In pure components, such free volume can be regarded as the molecular driving force behind the phenomenon called thermal expansion: a polymeric liquid expands when heated. Although such expansion is usually smaller than for low molecular weight liquids, its absolute value is sizable, see Table 4.1. Thermal expansivity is directly related to the Equation-of-State (EoS) behavior. An EoS is a relation between pressure (p), volume (V) and temperature (T) inside a liquid (or solid). See also Eq. (2.8). It is related to the free volume and, more fundamentally, to the free energy of the liquid. There is a

Table 4.1. Thermal expansivities of some polymers and solvents (from Patterson et al.[21])

	T, °C	Thermal expansivity, α, 10^{-4} K^{-1}
Polymer		
Polystyrene	135	5.8
Polybutadiene	25	6.9
Polyisobutylene	25	5.6
Polymethylmethacrylate	120	5.8
Polyvinylmethylether	135	7.2
Solvent		
Heptane	25	12.4
Butyl acetate	55	12.1
Carbon tetrachloride	20	12.7
Benzene	25	13.8

thermodynamic driving force for volume (or pressure) change upon temperature change at constant pressure (or constant volume).

Two polymeric melts will in general have different free volumes or thermal expansion coefficients (Table 4.1). Since a polymer chain in the melt is much more restricted in its degrees of freedom than a solvent, a polymeric liquid will have less free volume than a typical solvent and hence, the free volume difference between a polymer and its solvent will be large. A much smaller, but still significant, free volume difference between two polymeric liquids will in general exist because of differences in chain architecture and chain flexibility. Irrespective of constituent, two liquids of differing free volume will experience a net volume contraction upon mixing. Such contraction leads to an additional negative contribution to ΔH (favorable for mixing), and an additional positive contribution to $-T\Delta S$ (unfavorable for mixing). Recalling that $\Delta G_M = \Delta H - T\Delta S$, it follows that this additional contribution becomes more unfavorable with increasing temperature and ultimately leads to LCST demixing.

4.3.1.1 Helmholtz free energy

To understand the influence of free volume effects on the free energy of mixing more quantitatively, we need to use the Helmholtz free energy F. This is convenient, because F is coupled to the set of variables T, V, n_i, which is useful set when volume changes are to be taken into account. In addition, F is related to the partition function of the liquid. If we know one of the thermodynamic potentials as a function of the variables to which it corresponds, we can express all the other thermodynamic variables as a function of this potential and its derivatives. Hence, if we know ΔF, we also know ΔG, the latter remaining the thermodynamic potential relevant for describing the phase behavior. Recalling Eq. (2.6), the Helmholtz free energy F is defined by:

$$F = U - TS \qquad\qquad [4.10]$$

It is related to the Gibbs free energy G by

$$F = G - pV \qquad\qquad [4.11]$$

and the total differentials are related by:

$$dF = dG - pdV - Vdp \qquad\qquad [4.12]$$

F is related to the partition function Z by

$$F = -kT \ln Z \qquad\qquad [4.13]$$

If the partition function Z is known, all the thermodynamic properties of the system are known. This is what all EoS models do: to find (approximate) expressions for Z for the mixture and the components, hence to derive expressions for F and ΔF and, ultimately for ΔG to predict the phase behavior. The EoS is defined by:

$$p = kT \left(\frac{\partial \ln Z}{\partial V} \right)_T \qquad\qquad [4.14]$$

The tendency of a system to expand is governed by the partition function. In practice, it is impossible to derive an exact expression for this function for realistic systems. One works with approximate expressions (such as the FH combinatorial entropy of mixing) with some more or less empirical parameters (such as the interaction parameter χ). Hence, the concept "EoS model" bears relevance to the following:

- Some working expression for Z is derived or assumed.
- Pure component model parameters in Z are adjusted by fitting the experimentally determined EoS (pVT data, thermal expansion coefficients) of the pure components.
- The same expression for Z is applied in a relevant way to the mixture. Mixture model parameters are calculated from the pure components by applying simple mixing rules, o.a. based on the principle of corresponding states. In this way, an

expression for ΔG can be derived and the phase behavior can be predicted, in general by adjusting interaction parameters.

Flory[14] and Prigogine[15] have created the foundation to derive an analytical approximation for Z and hence, relevant expressions for ΔG, for mixing of polymeric species with non-negligible free-volume effects. It is assumed that the partition function can be separated into a part Z_{tr} representing the translational positions of the centers of mass of polymer segments and a part Z_i representing all other degrees of freedom, such as rotations and vibrations associated with the covalent bonds between polymer segments. These other degrees of freedom are generalized as "internal" degrees of freedom. We may write:

$$Z = Z_{tr}Z_i \qquad\qquad [4.15]$$

It is shown[15] that Eq. (4.15) is a good approximation as long as interaction forces are of the simple dispersive type. The important assumption is made that the free energy of mixing is determined by Z_{tr} exclusively, just as the intramolecular configurational entropy term cancels in the derivation of the FH entropy of mixing expression. Hence, it is implicitly assumed that there is no intramolecular configurational change upon mixing.

The classical expression for Z_{tr} for a system of N molecules is

$$Z_{tr} = \frac{1}{N!h^{3N}} \int dp_i \ldots \int dp_N \int dr_i \ldots \int dr_n e^{-\beta H} \qquad\qquad [4.16]$$

Due to the working of the Hamiltonian H ($H = \Sigma p_i^2/2m + U$), Z_{tr} splits up into the kinetic and the potential energy terms. The kinetic energy term depends on temperature exclusively and it is absorbed in Z_i (no influence on the free energy of mixing). The potential energy term is defined by the configurational partition function Q:

$$Q = \frac{1}{N!} \int dr_i \ldots \int dr_n e^{-\beta U} \qquad\qquad [4.17]$$

which may be extended to mixtures:

$$Q = \frac{1}{N_A!N_B!} \int dr_A^{N_A} dr_B^{N_B} e^{-\beta U} \qquad\qquad [4.18]$$

where $dr_i^{N_i}$ is an abbreviated notation for the coordinates of the centers of mass of the N_i molecules i.

Given the fact that Eqs. (4.15) and (4.16) hold, the EoS depends on Q exclusively:

$$p = kT\left(\frac{\partial \ln Q}{\partial V}\right)_T \qquad [4.19]$$

We can define the configurational free energy F_{conf}:

$$F_{conf} = -kT \ln Q \qquad [4.20]$$

Only F_{conf} appears in the free energy of mixing. Hence, without taking volume change upon mixing into account, we would have $pdV = Vdp = 0$, so that $\Delta F = \Delta F_{conf} = \Delta G_M$. In such case, the Helmholtz free energy of mixing equals the Gibbs free energy of mixing. Evaluation of the configurational partition function for athermal mixing of polymers would lead exactly to the same expression as the FH combinatorial entropy of mixing expression, along the same lattice approach outlined in detail in Chapter 3.

4.3.1.2 Corresponding states

Let us assume that the potential energy U is a sum of binary interactions only, and that such interactions can be represented by a universal function φ together with two scale factors r^* and ε^*:

$$\varepsilon(r) = \varepsilon^* \varphi \frac{r}{r^*} \qquad [4.21]$$

For one of the most used forms for dispersive forces, namely the 6–12 Lennard-Jones potential:

$$\varepsilon(r) = -\varepsilon^*\left(\frac{r^*}{r}\right)^6 + 2\varepsilon^*\left(\frac{r^*}{r}\right)^{12} \qquad [4.22]$$

the scale factors represent the coordinates of the minimum of $\varepsilon(r)$. The configurational integral, Eq. (4.17) can now be written in terms of reduced coordinates:

$$Q = \frac{r^{*3N}}{N!}\int d\left(\frac{r_1}{r^{*3}}\right)..\int d\left(\frac{r_N}{r^{*3}}\right)\exp\left[-\beta\varepsilon^*\sum_{i<j}\varphi\left(\frac{r_{ij}}{r^*}\right)\right] \qquad [4.23]$$

Apart from the factor r^{*3N}, Q depends only on the reduced quantities kT/ε^*, V/r^{*3} (integration) and N. We may write:

$$Q = r^{*3N}Q^*\left(\frac{kT}{\varepsilon^*},\frac{V}{r^{*3}},N\right) \qquad [4.24]$$

Since the configurational free energy F_{conf} (Eq. (4.20)) is an extensive variable, it must be of the form:

$$F_{conf} = Nf(T,v) \qquad\qquad\qquad [4.25]$$

where f depends only on the intensive variables T and v = V/N. We thus obtain

$$Q = r^{*3N}\left[q\left(\frac{kT}{\varepsilon^*},\frac{V}{r^{*3}}\right)\right]^N \qquad\qquad [4.26]$$

where q depends on intensive variables only. Defining the reduced variables

$$\overline{T} = \frac{kT}{\varepsilon^*},\ \ \overline{v} = \frac{v}{r^{*3}},\ \ \overline{p} = \frac{pr^{*3}}{\varepsilon^*} \qquad\qquad [4.27]$$

the EoS, Eq. (4.19) takes the universal form:

$$\overline{p} = \overline{p}(\overline{T},\overline{v}) \qquad\qquad\qquad [4.28]$$

Eq. (4.28) expresses the theorem of corresponding states. It implies that the pVT behavior of simple liquids is basically the same, when normalized on the molecular scale factors ε^* and r^*. The precise form of this function depends on the potential $\varepsilon(r)$ only. Exact expressions for the EoS of Lennard-Jones liquids or more simple hard sphere liquids can be found e.g. in Reference 15.

4.3.1.3 Cell partition function

In any liquid, also a polymeric liquid, a regularity in the spacing between neighboring molecules (segments) exists. This regularity is not as perfect as in a solid where long range translational order exists. The order in a liquid is rather short-range, with a sizable correlation between the positions of two neighboring molecules (segments), and hardly any correlation between molecules some 10 interatomic distances apart. Each molecule (segment) is confined to a "cell": a virtual cage in which the molecule can move freely. The size of this cage is determined by the relevant intermolecular interactions.

The cell concept was already introduced in the expression for the configurational partition function Q (Eq. (4.23)) and the corresponding universal EoS (Eq. (4.28)), because the molecules interacted via a Lennard-Jones potential. However, the contribution of the "cells" as such has not yet been identified explicitly. It is useful to define a cell partition function Ψ of a molecule (segment) in its cell referred to the energy of the particle at the center of the cell. The mean energy of interaction with neighboring particles $\omega(r)$ depends only on the distance r of the molecule (segment) from the center of the cell. We can write:

$$\Psi = 4\pi \int_{\text{cell}} \exp[-\beta(\omega(r) - \omega(0)]r^2 dr \qquad [4.29]$$

where $\omega(0)$ is the value of $\omega(r)$ at the center of the cell. The configurational partition function of a system of N molecules (segments) may now be written in the form

$$Q = \Psi^N \exp(-\beta N \omega(0)) \qquad [4.30]$$

where $N\omega(0)$ is the energy of the system when all particles are at the centers of their cells. The EoS, Eq. (4.19), can now be written as:

$$p = -\left(\frac{\partial \omega(0)}{\partial v}\right) + kT\left(\frac{\partial \log \Psi}{\partial v}\right)_T \qquad [4.31]$$

The pressure p arises from two contributions: a potential pressure generated by all molecules placed at the center of their cells and a thermal pressure resulting from the motions of the molecules in the cells. The volume v accessible to a molecule is a complicated function of the neighboring molecules and the exact intermolecular potential. Let v_0 be the hard core volume of the molecule. An exact expression for the cell partition function Ψ can be derived when a hard sphere interaction potential is assumed:[15]

$$\Psi = \frac{4\pi}{3}\gamma(v^{1/3} - v_0^{1/3})^3 \qquad [4.32]$$

where γ is a numerical factor which depends on the geometrical arrangement of the molecules. For a face centered cubic lattice, $\gamma = \sqrt{2}$. It can be shown that if the interaction potential is of the 6–12 Lennard-Jones form, the cell partition function is of the same form as Eq. (4.26). The hard core volume v_0 is related to the minimum volume v^* at the minimum of the Lennard-Jones potential by:[15]

$$v_0 = \frac{v^*}{2^{1/6}} \qquad [4.33]$$

4.3.1.4 Chain flexibility parameter

A relevant criterion for the existence of a lattice is that the mean distance between neighboring segments is approximately the same, whether such segments belong to the same polymer or not. Each polymer of chain length r has 3r degrees of freedom. Its Hamiltonian contains contributions from intramolecular as well as intermolecular interactions. Prigogine[15] assumed that these interactions are independent and separable. The lattice frequencies associated with intermolecular interactions are of the order 20 cm^{-1}, whereas the intramolecular frequencies (rotations, vibrations) are of the order of several 100 cm^{-1}. Hence, it is assumed that the 3r degrees of freedom of an r-mer can be separated into two groups:

- The internal degrees of freedom which depend only on intramolecular interactions.
- The external degrees of freedom which depend only on intermolecular interactions.

Again, it is assumed that only the external degrees of freedom contribute to the configurational partition function and therefore to the EoS and the free energy of mixing. For the total number of degrees of freedom, 3r, we have:

$$3r = n_e + n_i \qquad\qquad [4.34]$$

where n_e and n_i are the number of external and internal degrees of freedom, respectively.

The number of external degrees of freedom is given by

$$n_e = 3cr \qquad\qquad [4.35]$$

Equation (4.35) defines the chain flexibility parameter c (or Prigogine c parameter; $0 \le c \le 1$). It expresses the connectivity of an r-mer:

- $c = 0$: The situation where $n_e = 0$ and all degrees of freedom are internal: the fully rigid chain.
- $c = 1$: No connectivity at all. The behavior is identical to that of r separate monomers: the fully flexible chain.
- $0 < c < 1$: Realistic chain with more (c→1) or minor (c → 0) degree of flexibility.

4.3.2 THE FLORY EoS THEORY

The key concepts outlined above are the foundations of EoS thermodynamics. Such theory is historically associated with the names of Flory,[14] Prigogine,[15] and of Flory together with various coworkers, in particular Orwoll and Vrij.[16-19] Let us again consider the mixing of N_i polymers of species i (i = 1,2 for binary mixtures) with chain lengths r_i. The configurational partition function Q can be written in terms of the cell partition function Ψ of the respective segments and the lattice energy of all segments at their equilibrium positions. In the pure components, the total number of cells amounts to $N_i r_i$. If the segments were to be considered as uncoupled molecules rather than as connected into a chain, the configurational partition function would read:

$$Q = \frac{4\pi}{3}\gamma_i(v_i^{\frac{1}{3}} - v_0^{\frac{1}{3}})^{3N_i r_i} \exp(-\beta E_i(0)) \qquad\qquad [4.36]$$

where γ_i is the numerical constant of component i that depends on the coordination of the lattice. $E_i(0)$ is the lattice energy of all segments at their equilibrium positions. The cell partition function in eq. 4.36 is raised to the power $3N_i r_i$, which may be interpreted as bearing relevance to the $3r_i$ degrees of freedom of N_i large molecules. Due to the fact that we need to consider N_i polymers with a certain chain connectivity, rather than N_i large molecules, Flory proposed that the number of degrees of freedom becomes $3c_i r_i$ rather than $3r_i$. The configurational partition function of a system of N_i polymers with chain flexibility parameter c_i thus becomes:

$$Q = \frac{4\pi}{3}\gamma_i(v_1^{\frac{1}{3}} - v_0^{\frac{1}{3}})^{3c_i N_i r_i} \exp(-\beta E_i(0))$$
[4.37]

For a binary mixture, the configurational partition function reads:

$$Q = \Omega(N_1, N_2)\frac{4\pi}{3}\gamma_M(v_M^{\frac{1}{3}} - v_{M,o}^{\frac{1}{3}})^{3c_M N r_M} \exp(-\beta E_M(0))$$
[4.38]

$\Omega(N_1, N_2)$ is the normal FH combinatorial factor of mixing two polymers on a lattice, Eq. (3.31). In the evaluation of the free energy of mixing, ΔF, this term leads to the FH entropy of mixing. It is also evaluated under the assumption that there is no intramolecular configurational entropy change upon mixing. The γ_M, v_M, $N(N_1 + N_2)$, c_M, r_M, and E_M represent the relevant average values for the respective parameters in the mixture. They are either predicted from appropriate averaging rules or fitted by adjusting experimental phase diagrams or pVT data of the mixture. It is mainly in the theoretical treatment of these parameters that various "generalized" versions of the original Flory model differ.

The Flory model, and generalizations of it, have been reviewed by McMaster.[20] Instead of deriving the lattice energy from an assumed interaction energy, it is more simple to take a phenomenological dependence on cell volume:

$$E_i(0) \sim \frac{1}{v^n}$$
[4.39]

where n is an adjustable parameter, expected within a range between 1 and 1.5, based on energy of vaporization data of small molecular weight analogues, in correspondence with the definition of the solubility parameter (Chapter 2.5). The following pure component EoS is obtained:

$$\frac{\bar{P}_i \bar{v}_i}{\bar{T}_i} = \frac{\bar{v}_i^{\frac{1}{3}}}{\bar{v}_i^{\frac{1}{3}} - 1} - \frac{1}{\bar{T}_i \bar{v}_i^n}$$
[4.40]

where

$$\overline{P}_i = \frac{P}{P_i^*} \qquad\qquad [4.41]$$

$$\overline{T}_i = \frac{T}{T_i^*} \qquad\qquad [4.42]$$

$$\overline{v}_i = \frac{v}{v_i^*} \qquad\qquad [4.43]$$

The starred quantities are parameters to be fitted from the pure component pVT behavior. They are defined by Eq. (4.27). The EoS for the mixture can be written in a form identical to Eq. (4.40), if the reduction parameters P^* and T^* are defined as relevant averages of the pure component parameters, and if a cross-interactional energy parameter χ, similar to the FH interaction parameter, is added.

For the most primitive Flory model, the remaining mixture parameters follow straightforwardly from the pure components:

$$c_M = \Phi_1 c_1 + \Phi_2 c_2 \qquad\qquad [4.44]$$

$$r_M = \frac{N_1}{N} r_1 + \frac{N_2}{N} r_2 \qquad\qquad [4.45]$$

$$\gamma_M = \gamma_1 = \gamma_2 \qquad\qquad [4.46]$$

McMaster showed that relevant expressions for the free energy of mixing, and the chemical potentials of each component can be obtained so that binodals and spinodals can be computed. Using EoS parameters of polystyrene and polyethylene, and setting the interaction parameter χ to zero, the EoS effects could be studied separately. It is shown that in this way LCST demixing is obtained. Due to the normal working of the FH combinatorial entropy of mixing term, the binodal curve shifts to higher temperatures as the molecular weight of either of the two components decreases (increasing miscibility), and the binodal curve becomes more skewed as the difference in chain lengths between the two components increases. It is also shown that the free volume v_M of the homogeneous mixture is always less than that of the demixed system when the free volumes of the pure components are different ($v_1 \neq v_2$). In other words, LCST demixing goes together with volume contraction.

If, in addition, a finite value for the interaction parameter χ is taken, simultaneous UCST and LCST behavior is obtained. In general, a homogeneous one-phase region between UCST and LCST is shown to exist. If either the free volume difference between the pure components, or the value of the interaction parameter, is increased, an hour-glass shaped phase diagram appears, at the expense of a vanishing one-phase region.

Although the approach of McMaster does show the general LCST behavior, the relevant equations for the binodals and spinodals are rather complicated. Alternatively, the order of magnitude of free volume effects can be estimated using a simplified version of Flory's model, proposed by Patterson et al.[21] All free volume contributions to the free energy of mixing are lumped together in the the FH interaction parameter χ. The interaction parameter is split-up into two terms:

$$\chi = \chi_{disp} + \chi_{free} \qquad [4.47]$$

where χ_{disp} is the purely dispersive part of the interaction parameter (Eq. (4.2)). χ_{free} is defined by:

$$\chi_{free} = \frac{c \overline{v}_1^{1/3}}{2\left(\frac{4}{3} - \overline{v}_1^{1/3}\right)} \tau^2 \qquad [4.48]$$

where c is the chain flexibility parameter, taken equal for pure components and mixture. The reduced volume of the pure components is simply related to their thermal expansivity α:

$$\overline{v}_i^{1/3} = 1 + \left(\frac{\alpha_i T}{3(1 + \alpha_i T)}\right) \qquad [4.49]$$

The characteristic free volume difference τ is defined by

$$\tau = 1 - \frac{\overline{T}_2}{\overline{T}_1} \qquad [4.50]$$

The reduced temperatures of component 1 and 2, respectively, are related to the reduced volumes by the EoS:

$$\overline{T} = \frac{\overline{v}^{1/3} - 1}{\overline{v}^{4/3}} \qquad [4.51]$$

which was derived from the Flory EoS, Eq. (4.40), under the condition n = 1 and reduced pressure P = 0 (no influence of pressure). The temperature dependence of the free volume term χ_{free} can now easily be calculated from the difference in thermal expansivities of the pure components. Using $\alpha_1 = 5.10^{-4} \, K^{-1}$ and $\alpha_2 = 6.10^{-4} \, K^{-1}$ (already a sizable difference for polymers, see Table 4.1), χ_{free} is predicted as shown in Figure 4.1. It is positive (unfavorable for mixing) at any temperature,

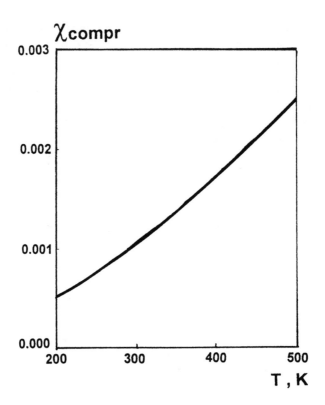

Figure 4.1. χ_{free} as a function of temperature. (Adapted, by permission, from A. Wakker and M. A. van Dijk, *Polym. Networks & Blends*, **2**, 123 (1992)).

increasing smoothly with temperature. LCST demixing occurs when χ_{free} exceeds the critical value χ_c defined by Eq. (3.59).

It should be emphasized that although Flory EoS theory does predict the LCST behavior, together with UCST demixing when a finite value for the dispersive interaction parameter is taken, none of the existing models is fully satisfactory in describing experimental phase diagrams. EoS models need the same approximations for the combinatorial entropy of mixing as originally employed by Flory and Huggins. Such entropy of mixing still makes a sizable contribution to the free energy of mixing, although it is a very small term as such for polymeric mixtures. In addition, so many subtle physical contributions from interaction energies and free volume effects contribute to the free energy of mixing. Last but not least, actual compressibility effects in mixtures are extremely sensitive to differences in pure component EoS parameters, e.g., thermal expansion coefficients. EoS models describing the phase behavior of mixtures could benefit from a more accurate description of the EoS behavior of pure components.

Such attempt was made by Walsh and Dee.[22,23] Based on theory developed by Flory and Eichinger,[18] an extra phenomenological term Q is added to the cell partition function, Eq. (4.32):

$$\Psi = \frac{4\pi}{3}\gamma \, (v^{V_3} - Qv_0^{V_3})^3 \qquad\qquad [4.52]$$

In cell models, the cell partition function strongly depends on the interaction potential coupled with the particular geometry employed. Therefore, the identification of a putative hard-core cell volume v_0 is not trivial. The introduction of an extra adjustable parameter Q generates somewhat more flexibility to describe the local geometry of a closed-packed polymeric liquid. The drawback is that if this modification is physically meaningful, a unique value for Q should be found irrespective of the particular EoS behavior of various polymeric liquids.

This has been investigated, again by assuming an FCC lattice ($\gamma = \sqrt{2}$). It was found that the EoS behavior of various polymers can be described more accurately with the modified cell model than with traditional cell models. A unique value of $Q = 1.07 \pm 0.02$, which provides the optimum fit irrespective of polymer, was obtained. Such value corresponds with an approximately 25% increase in the effective hard-core volume of the cell model. In other words, in reality there appeared to be less free volume available for polymer segments than expected from the potentials used.

4.3.3 HOLE MODELS

4.3.3.1 Key concepts

A special class of EoS models, of which the results are comparable with those obtained from Flory EoS models, are the so-called hole models. Such models are historically associated with the names of Sanchez and Lacombe (SL),[24-28] and Koningsveld and Kleintjes (KK).[9-11] Methodology employed by both schools is basically equal. The difference mainly comes from the fact that for the description of the interaction parameter χ, KK employ the phenomenological approach, in particular with regard to the concentration dependence of the interaction parameter, as outlined in Chapter 4.2.

In hole models, compressibility is modeled by allowing empty sites (holes) on a lattice filled with r-mers, of which every mer has a fixed volume v^*. There is no cell free volume. The available free volume is exclusively determined by the fraction of empty sites. Such theories are also known as lattice fluid (LF) theories. Interaction energies between holes and holes or holes and segments are set to zero. Both

Flory and hole theories require three EoS parameters for each pure component.

The concept behind hole models is that pressure and temperature changes cause variations in the concentrations of holes, whilst the volume per lattice site is constant. Unlike Flory models, neither sophisticated expressions for cell partition functions, nor chain flexibility parameters, are needed. One can straightforwardly derive expressions for the Helmholtz free energy using normal FH combinatorics for placing N molecules of chain length r and N_o empty sites on a lattice with $N_o + rN$ sites in total. The same approximations as in the derivation of the FH combinatorial entropy of mixing are used. The relevant expressions for the entropy of mixing ΔS_M reads:

$$\Delta S_M = \Delta S_{comb} + \Delta S_{holes} \qquad [4.53]$$

where ΔS_{comb} is the usual Flory-Huggins combinatorial entropy of mixing term (per mole of molecules)

$$\frac{\Delta S_{comb}}{R} = \frac{\Delta S_{FH}}{R} = \frac{\Phi_1}{r_1} \ln \Phi_1 + \frac{\Phi_2}{r_2} \ln \Phi_2 \qquad [4.54]$$

and ΔS_{holes} is the entropy of mixing holes with the two (!) polymers:

$$\frac{\Delta S_{holes}}{R} = \frac{(1 - \bar{\rho})}{\bar{\rho}} \ln (1 - \bar{\rho}) + \frac{\ln \bar{\rho}}{r_M} \qquad [4.55]$$

where $\bar{\rho}$ is the reduced density of the mixture, which equals the fraction of occupied sites, and $1 - \bar{\rho}$ is the fraction of vacant sites (holes). The average chain length r_M is defined by:
For the pure components, the free energy F is at a minimum and satisfies the following EoS ($p = -\partial F / \partial v$):

$$\frac{1}{r_M} = \frac{\Phi_1}{r_1} + \frac{\Phi_2}{r_2} \qquad [4.56]$$

$$\frac{1}{\bar{v}^2} + \bar{p} + \bar{T} \left[\ln \left(1 - \frac{1}{\bar{v}} \right) + \left(1 - \frac{1}{r} \right) \frac{1}{\bar{v}} \right] = 0 \qquad [4.57]$$

where the reduced volume v is defined as :

$$\bar{v} = \frac{1}{\bar{\rho}} \qquad [4.58]$$

EoS parameters p^*, V^* and T^* can be determined from a fit to experimental density data.

The power of hole models is that polymeric mixtures are treated as three component systems (two polymeric species and holes). This leads to relatively simple expressions for the free energy of mixing. In addition, relatively simple mixing rules for p^*, V^*, and T^* can be used. The mixing rules all refer to the pure component properties of the corresponding closed packed system (reduced density $\rho = 1$). In the SL model, the interaction parameter can in principle be predicted from such pure component properties, whereas in the KK model an adjustable phenomenological form is taken.

The general result of hole theories is that differences in EoS properties of the pure components make a thermodynamically unfavorable contribution to the free energy of mixing. This is most apparent from the spinodal condition, Eq. (3.56), when it is derived when allowance is made for empty sites on the lattice:[27]

$$\frac{1}{r_1\Phi_1} + \frac{1}{r_2\Phi_2} > 2\bar{\rho}\left[\chi + \frac{1}{2}\Psi^2\bar{T}p^*\beta\right] \qquad [4.59]$$

where the left hand side of Eq. (4.59) comes from the FH combinatorial entropy of mixing, $\bar{\rho}\chi$ is the interaction parameter contribution and $\bar{\rho}\Psi^2Tp^*\beta$ is an entropic contribution from compressibility effects. It is a positive term, making an unfavorable contribution to the spinodal and hence, its presence does not favor miscibility. Its working is similar to the term χ_{free} obtained with Patterson's adaptation of the Flory model (Eq. (4.48)), and LCST demixing is obtained. In addition, the term $\bar{\rho}\chi$ has the usual T^{-1} dependence and leads to UCST demixng. It makes a positive (unfavorable) contribution to the enthalpy of mixing.

A schematic representation of the three terms in the spinodal condition, Eq. (4.59), is given in Figure 4.2. The parameter β represents the isothermal compressibility of the mixture, p^* is the characteristic pressure of the mixture, T is the reduced temperature of the mixture, and Ψ is a function of the pure component parameter differences, especially T^* values, and is in general non-zero. Like in the Flory theory, it is the difference in pure component EoS parameters that leads to LCST demixing. Such difference leads to volume contraction, and contributes favorably to the enthalpy of mixing.

In Flory-type models, LCST demixing is modeled by cell free volume, in hole models by empty sites. The effective result is the same. Possibly, hole models bear less actual relevance to real polymeric systems, due to the fact that a minimum of statistical mechanics is used to take chain connectivity into account. There is no chain flexibility parameter in hole models. The quantitative fit to experimental phase diagrams is in general not so good. Both hole and Flory models employ approximate EoS descriptions of the pure components. Better

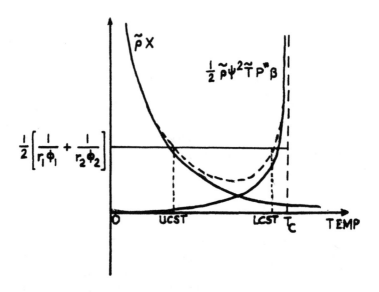

Figure 4.2. Schematic behavior of the three terms in the spinodal condition, Eq. (4.59), as a function of temperature. Dashed curve is the sum of the two terms on the rhs. (Adapted, by permission, from I. C. Sanchez and R. H. Lacombe, *Macromolecules*, **6**, 1145 (1978)).

EoS descriptions of the components could produce better descriptions of the phase behavior of actual mixtures.

4.3.3.2 Holes and Huggins

The most general EoS models are those that allow for both cell free volume and lattice vacancies. This results in a cell free volume that depends on environment, which is a mixture of occupied cells (segments) and holes. This has originally been done in the Simha-Somcynsky (S&S) theory.[29,30] A modification of this model, the "Holes and Huggins" (HH) model, was developed by Nies and coworkers.[31-33] HH differs from S&S mainly in the sense that for the calculation of the combinatorial entropy of mixing the Huggins correction has been added, which should in principle raise the predictive power.

The drawback however is that the cell partition function, through the interaction potential, now becomes a relatively complex function of the specific lattice used and the fraction of empty sites. Therefore, in the HH model, it is assumed that segments interact via the 6–12 Lennard-Jones potential on a lattice with a fixed coordination number z of 12. This corresponds with a face centered cubic (FCC) lattice, resembling a close packing of spheres. Such packing, together

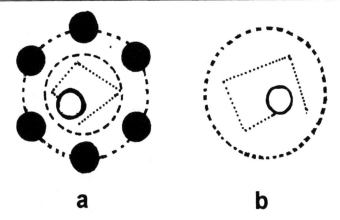

Figure 4.3. HH solid (A) and gas (B) like cells. Adapted, by permission from E. Nies and A. Stroeks, *Macromolecules*, **23**, 4088 (1990).

with the possibility of creating holes in it and of relaxing the cell volume, is assumed to represent something like a disordered liquid.

Exact expressions for the cell partition function can be derived assuming that either all adjacent sites are occupied ("solid-like", Figure 4.3 A) or all adjacent sites are empty ("gas like", Figure 4.3 B). The actual partition function can be fitted by taking a relevant average over the two extreme conditions. Again, by defining simple mixing rules, the number of adjustable parameters in the model can be minimized.

The attractiveness of the HH model is that due to fact that all possible physical features are incorporated in the theory, more complicated experimental data, such as pressure dependence of binodals, volume contraction, and enthalpy of mixing, can be predicted qualitatively.

4.4 SPECIFIC INTERACTION MODELS

4.4.1 KEY CONCEPTS

LCST phase separation can be caused by specific interactions as well. In polymer blends where specific interactions play a role, LCST demixing is expected to be dominated by such interactions rather than by compressibility effects, because of the relatively small difference in thermal expansivity between polymers. Unlike for compressibility effects, specific interactions can cause interaction parameters to become negative.

In contrast to dispersive interactions, where heat is consumed upon mixing, specific interactions give rise to a heat release upon mixing and hence, a negative (favorable) contribution to the free energy of mixing. Specific interactions are caused by synergistic interaction forces between unlike polymer segments such that the resulting interaction is stronger than the average of like-like interactions. Such interactions include hydrogen bonding and Lewis acid-base interactions.[34] Although specific interactions are favored enthalpically, they are entropically unfavorable because degrees of freedom are "frozen in" in one specific interaction channel. This leads to an additional entropy loss term in the free energy of mixing, which is positive. This unfavorable entropy loss becomes more important with increasing temperature and leads to LCST demixing.

The difficulty in incorporating the effect of specific interactions in the free energy of mixing lies in how to quantify the local ordering effects associated with the formation of such interactions. On a molecular level, the basis for such quantification of non-random mixing effects has been developed by Guggenheim, the so-called quasi-chemical theory.[35] The relevant expressions for the net effect on the free energy of mixing are complicated and it is a tough task to associate them with the phase behavior of realistic molecular liquids, let alone polymeric liquids. The relevant theory behind the prediction of the phase behavior of polymeric mixtures in which specific interactions play a role is therefore relatively less mature than that behind EoS phenomena.

We will discuss here a version of the Guggenheim quasi-chemical theory, adapted for polymeric mixtures, developed by Ten Brinke et al.[36] This model contains the essential physical ingredients to understand why specific interactions lead to LCST demixing.

Two different monomers A and B can form a specific interaction with energy $U_1 < 0$, or a dispersive interaction with energy $U_2 > 0$. It is assumed that a specific interaction is formed only if the monomers are in the same state in configuration space (for instance, a certain orientation). If there are q different states available for each monomer, then a specific interaction can only be formed in q ways, whereas a dispersive interaction can be formed in $q(q-1)$ ways: the more degrees of freedom a monomer has, the higher the price that has to paid for forming a specific interaction.

Within the framework of the original FH formulation, the usual combinatorial entropy of mixing expression is retained, and all other contributions to the free energy of mixing (both enthalpic and entropic) are lumped together in the interaction parameter. The expression for the total interaction parameter, χ, excluding possible compressibility effects, becomes:

$$\chi(T) = \chi_{disp}(T) + \chi_{spec}(T) = \chi(z, q, U_1, U_2, T) =$$

$$= z \left[\frac{1}{RT}U_2 + \ln (1 - \lambda) + \ln \left(\frac{q + 1}{q} \right) \right] \qquad [4.60]$$

where z is a lattice coordination number and λ is the fraction of directional specific interactions, defined by

$$\lambda = \left[1 + q \exp\left(\frac{U_1 - U_2}{RT} \right) \right]^{-1} \qquad [4.61]$$

Note that χ has the property that

$$\lim_{T \to \infty} \chi(T) = 0 \qquad [4.62]$$

We can separate the interaction parameter χ into an enthalpic component χ_h and an entropic component χ_s according to:

$$\chi_h = -\frac{\partial \chi}{\partial T} \qquad [4.63]$$

$$\chi_s = \frac{\partial (T\chi)}{\partial T}$$

For this model, the two components are given by

$$\chi_h = \left(\frac{z}{RT} \right)(\lambda U_1 + (1 - \lambda)U_2) \equiv (\chi_h)_{spec} + (\chi_h)_{disp} \qquad [4.64]$$

$$\chi_s = z \left[\ln (1 - \lambda) + \ln \left(\frac{q + 1}{q} \right) - \frac{\lambda(U_1 - U_2)}{RT} \right] \qquad [4.65]$$

We defined in Eq. (4.64) a specific enthalpic interaction part $(\chi_h)_{spec} = (z/RT)\lambda U_1$ (proportional to λ, the fraction specific interactions) and a dispersive enthalpic interaction part $(\chi_h)_{disp} = (z/RT)(1 - \lambda)U_2$ (proportional to $1 - \lambda$). The working of the model is illustrated in Figure 4.4, where χ, χ_h and χ_s are calculated as a function of temperature for the set of parameters $(U_1, U_2, q, z) = (-0.75$ kcal/mol, 0.1 kcal/mol, 15, 4). Using the solubility parameter concept for dispersive forces:

$$U_2 = V_{lat}(\Delta\delta)^2 \qquad [4.66]$$

and a typical value for V_{lat} of 100 cm^3/mol, the value for U_1 corresponds to a solubility parameter difference $\Delta\delta$ of 1 $\sqrt{(cal/cm^3)}$.

For this set of parameters, χ is negative for $T < 335K$, so that a high molecular weight polymer blend will be completely immiscible for $T < 335K$. Above this temperature, LCST demixing into two phases

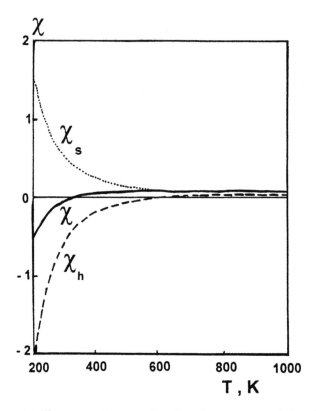

Figure 4.4. The χ, χ_s and χ_h as a function of temperature. (Adapted, by permission, from A. Wakker and M. A. van Dijk, *Polym. Networks & Blends*, **2**, 123 (1992)).

with different compositions will occur. Because the shape of the coexistence curve of polymer blends is usually rather flat, the mixture will still be partially miscible just above 335K, but immiscible at higher temperatures.

Figure 4.4 also shows the entropically driven character of lower critical demixing: At low temperature, χ_h is negative and in absolute value larger than χ_s, so that a stable, miscible mixture results. At higher temperatures however, the larger entropy penalty drives the mixture towards immiscibility.

Typical compressibility effects (Figure 4.1) are much smaller than effects of specific interactions. Moreover, unlike specific interactions, such effects do not lead to negative interaction parameters. From such first, rough comparison, it is expected that if a polymer mixture shows LCST demixing, there is good chance that specific interactions dominate the phase behavior. This is even more likely when UCST demixing is not observed.

The implication for predicting polymer–polymer miscibility is that the temperature T_c, where $\chi_c = \chi(T_c) = 0$, is determined by the competition between dispersive forces and specific interactions: The larger U_2 (and hence, the larger χ_{disp}), the lower T_c; the larger q (and hence, the larger χ_s), the lower T_c; the larger U_1 (and hence, the larger χ_{spec}), the higher T_c.

In order to get quantitative insight into how U_1, U_2, and q influence the location of the critical condition $\chi_c = \chi(T_c) = 0$, the problem $\chi(q,U_1,U_2,T_c) = 0$, which is independent of z, needs to be solved. Such result is presented in Figure 4.5, where $\Delta\delta$ is shown as a function of U_1, at fixed values $T_c = 500K$ and $q = 15$.

The $\chi = 0$ curve approaches asymptotically a certain value of $|U_1|$. Beyond this value, the system is always miscible, irrespective of $\Delta\delta$. This "phase transition" can be understood as follows: If $|U_1|$ increases, the fraction of specific interactions λ increases, until all interaction channels are occupied: $\lambda = 1$. In the limit $\lambda \to 1$, we have $q.\exp((U_1 - U_2)/RT) \ll 1$, so that

$$\lambda = \frac{1}{1 + q\, \exp\left(\dfrac{U_1 - U_2}{RT}\right)} = 1 - q\, \exp\left(\frac{U_1 - U_2}{RT}\right) \qquad [4.67]$$

Hence

$$\lim_{\lambda \to 1} \chi/z = \frac{U_2}{RT} + \ln(q) + \frac{(U_1 - U_2)}{RT} + \ln\left(\frac{q+1}{q}\right) = \frac{U_1}{RT} + \ln(q+1) \; [4.68]$$

which is independent of U_2! There is a limiting value $U_1 < -RT_c \ln(q+1)$ (for our model mixture, -2.74 kcal/mol, Figure 4.5, dashed line), for which there is complete miscibility independent of U_2, because all specific interaction channels are occupied.

The dot in Figure 4.5 denotes the position of our model mixture in "interaction space". Evidently, the mixture can be made miscible either by increasing $|U_1|$, decreasing $\Delta\delta$, or decreasing q (not shown).

In practice, both U_1 and q will be fixed by the chemical nature of the specific interaction between two polymers: A hydrogen bond may be strong or weak, more or less directional specific, etc. The solubility parameter difference however, is in general not fixed: it can be influenced easily by adding simple groups (-CH_2-, -CH_3, -O-) to either polymer. Consequently, polymer-polymer miscibility can be brought about by:

- Ensuring that specific interactions between polymer A and B segments are present.

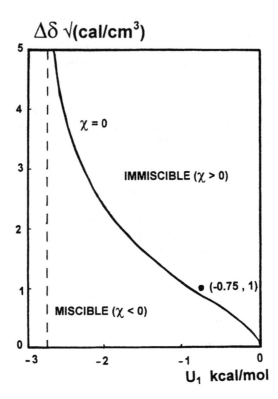

Figure 4.5. The $\Delta\delta$ as a function of U_1 under the condition $\chi = 0$ and $q = 15$ (the dot denotes the corresponding location of the model mixture. (Adapted, by permission, from A. Wakker and M. A. van Dijk, *Polym. Networks & Blends*, **2**, 123 (1992)).

- Limiting the solubility parameter difference by means of the non-specific interaction groups, such that miscibility at a certain processing temperature is ensured.

The influence of the choice of the critical temperature T_c on the competition between dispersive and specific interactions is shown in Figure 4.6. The model mixture starts to become miscible between 300 and 500K, which is evident from a comparison with Figure 4.4. The larger $(-U_1, \Delta\delta)$, the stronger the temperature difference. At low values of $(-U_1, \Delta\delta)$, the temperature dependence vanishes, and $\partial\chi_c/\partial T \to 0$. This is the same effect as the flattening of the $\chi(T)$ curve in Figure 4.4. It shifts towards lower temperatures when $(-U_1, \Delta\delta)$ decreases. Hence, in miscible mixtures in which $|U_1|$ is small, $\Delta\delta$ must also be small, and no correlation between solubility parameter difference and the demixing temperature T_c exists: The mixture is either completely miscible, or not.

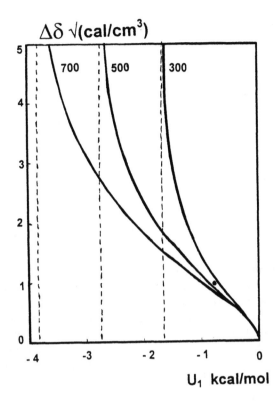

$\Delta\delta \sqrt{(cal/cm^3)}$

U_1 kcal/mol

Figure 4.6. Temperature dependence of the critical $\chi = 0$ curve, for $q = 15$. (Adapted, by permission, from A. Wakker and M. A. van Dijk, *Polym. Networks & Blends*, **2**, 123 (1992)).

4.4.2 INTERACTION STRENGTHS

From the preceding section it has become clear that basically polymer-polymer miscibility is a matter of competition between simple dispersion forces and specific interactions. EoS effects may further complicate matters, but will always have a destabilizing effect and never lead to negative interaction parameters. Although the most general thermodynamic models will need to take into account dispersion forces, specific interactions, and compressibility effects as well, let us assume here that the influence on LCST behavior of specific interactions dominates the influence of compressibility effects.

In such case, a useful description of the phase behavior of polymeric systems in which specific interactions play a role depends on the ability to separate simple dispersive forces from specific interaction forces. In other words, what is needed is a useful prediction of the pure dispersive part of interaction energies of model chemical groups involved in a directional specific interaction. Some guidelines

Table 4.2. Allowed solubility parameter difference $\Delta\delta$ given a certain specific interaction strength (from Coleman et al.[38]).

Specific interactions involved	$(\Delta\delta)_{crit}$, $\sqrt{(cal/cm^3)}$
Dispersive forces only	0.1
Dipole-dipole	0.5
Weak	1.0
Moderate	1.5
Moderate to strong	2.0
Strong	2.5
Very strong	3.0

are given by Coleman, Painter and coworkers.[38] For purely dispersive interactions, it is confirmed that Small's group contribution scheme is still the most consistent one, provided that not only molar interaction energies, but also group contributions to the molar volumes based on the same set of model compounds, are employed. In this way, when compared to relevant experimental data, the standard error involved in the prediction of solubility parameters of non-polar polymers amounts to 0.2 $\sqrt{(cal/cm^3)}$.

It is argued that the dispersive solubility parameter of specific interacting groups can be predicted to a reasonable extent as long as molar group contributions from non- or weakly specific interacting model compounds are available. For instance, the dispersive solubility parameter of an amine group, -(C=O)-(N-H)-, should be considered as the combination of separate C=O and N-H contributions, derived from non-specific interacting model compounds. The same arguments hold for polyurethanes. In case of stronger specific interactions, in particular hydrogen bonds, the less worse strategy appears to be to completely eliminate the delocalized hydrogen atom in the hydroxyl (O-H) group, and to employ the ether (-O-) group contribution exclusively. In this way, a consistent solubility parameter data set of almost all polymers, based on dispersive interactions exclusively, was established. The interested reader is encouraged to read the book of Coleman, Graf and Painter [38] for more specific details.

Based upon some quantitative insight into specific interaction strengths and actual phase diagrams of miscible polymeric systems, a qualitative, empirical categorization of the miscibility of high molecular weight polymer mixtures was obtained, see Table 4.2. In this table, the allowed solubility parameter difference, given a certain interaction strength, in order to obtain a miscible system, is shown.

It should be noted that the error in estimating solubility parameters increases with increasing interaction strength, because of increasing difficulty in separating both forces.

In order to make more quantitative progress in this area, it will be necessary to predict specific interaction strengths as well as the inevitable ordering effects associated with it, by independent means. Spectroscopic methods are available to determine interactions strengths. It is desirable to measure such interactions in the actual polymer melt. Reason for this is that due to the covalent bonding of functional groups in polymers, such groups are more restricted in their degrees of freedom than their low molecular weight analogues. Hence, interaction strengths, as well as associated ordering effects, are not expected to be the same a priori.

4.4.3 GENERALIZATIONS

The most general model that one needs is a model able to predict the phase behavior of any polymeric mixture, that is a mixture in which simple dispersive interactions, compressibility effects as well as specific interactions are allowed to play a role. Both Flory type EoS models and hole models can in principle be modified to allow for specific interactions as well. To date, the most successful approach appears to be the one of Ten Brinke, because the game is played on a lattice, there is no complicated cell partition function, and all essential specific interaction contributions are lumped together into the interaction parameter. Sanchez and Balasz[39] have thus generalized the original Sanchez-Lacombe hole model in a consistent way, which resulted essentially in a compressible version of the Ten Brinke model. The free energy of mixing expression reads:

$$\frac{\Delta F}{RT} = \frac{\Delta S_{comb}}{R} + \frac{\Delta S_{holes}}{R} + \bar{\rho}\chi\Phi_1\Phi_2 \qquad [4.69]$$

where ΔS_{comb} and ΔS_{holes} are defined by Eqs. (4.54) and (4.55), respectively, and $\bar{\rho}$ is the reduced density of the mixture. The interaction parameter χ equals the Ten Brinke specific interaction parameter defined in Eq. (4.60). Note that if $\bar{\rho}$ is set to unity (no holes on the lattice), we have $\Delta S_{holes} = 0$ and the Ten Brinke free energy of mixing expression is recovered.

The generalized model thus obtained has been compared with neutron scattering data from the model mixture polystyrene (PS) – poly vinyl methyl ether (PVME), in which negative interaction parameters have been observed. The mixture has weakly polar interactions, and it shows LCST demixing in a relevant temperature region. It is shown that thanks to the introduction of the specific interaction parameters U_1 and q, a relatively accurate description of the temperature dependence of the interaction parameter is obtained. On the other hand, the actual concentration dependence of the spinodal is determined exclusively by the chain lengths (combinatorial entropy of mixing) and by compressibility effects. The actual fit is especially sensitive to the choice of pure component parameters T_i^*. It does not depend on specific interaction parameters whatsoever. In addition, at the LCST critical point, the absolute contribution to the interaction parameter of compressibility effects is sizable. The results show how various subtle effects govern the phase behavior of this mixture. Model results will discussed in more detail in Chapter 6.

4.5 THE (n=0) VECTOR MODEL

The Flory-Huggins theory is often referred to as a lattice theory. This is only partly true. Strictly speaking, the only use of a lattice that is made is in designating a fixed number of neighbors z to each polymer or solvent segment and in prescribing a fixed amount of possible segment positions. The longer range lattice topology does not play a role. For example, the fact that turning left 3 times on a 2 dimensional simple cubic lattice brings one back to the origin is not included in FH theory. In fact, one could say that the Flory-Huggins lattice is the very special type of lattice where no return to previous positions is possible. Such a lattice is known as a Bethe lattice alias Caley tree. Figure 4.7 shows the 2 dimensional simple cubic equivalent Bethe lattice. It is known that calculations on Bethe lattices reproduce the results of mean field theory which indeed the FH theory is. Calculations on "normal" connected lattices are much more difficult due to the long range connectivity.

One could argue that calculations on lattices are not very interesting for practical problems because 1: different lattices produce different results and 2: the real world is a continuum anyway. Such an argument cannot be nullified but may be weakened by noting that for many problems the exact choice of the lattice or the use of a continuum is irrelevant. For example, the end-to-end distance of a random walk will always scale with the square root of the number of steps irrespective of lattice choice or use of a continuum. The behavior of many parameters near a critical point is independent of the choice of the particular lattice and is described by universal critical exponents that only depend on the dimensionality of the problem.

Nevertheless, for exact numerical calculations far away from critical points and other extremities, where all the nitty gritty of the model plays its role, the choice of the lattice is reflected in the results. Usually it is manifested by one or more lattice parameters that are used as empirical fit parameters for the comparison with experimental results.

Figure 4.7. Two-dimensional "simple cubic" Bethe lattice.

A Flory-Huggins type model of a polymer solution on a 'true' lattice requires chains to be treated as self-avoiding walks which is an extremely difficult problem, not only because of the long range connectivity of the lattice, but also because one has to keep track of individual chain configurations. It has been shown that the latter aspect of the problem may be cured by an ingenious trick. The problem of calculating the partition function of self-avoiding walks on a lattice is mathematically equivalent to the seemingly unrelated problem of n-dimensional magnetic spins on a lattice in the seemingly absurd limit of n=0. The spin problem does not inherently contain the task of keeping track of individual chains.

We will now briefly outline the principles of the theory, which is rather involved in its details, in order to understand the basic trick and appreciate the building of theoretical developments that has been constructed to exploit this idea in the area of polymer thermodynamics. The theory has been introduced by de Gennes[40,41] and extended by Freed[42] and coworkers.

Consider a lattice with a magnetic spins $\vec{S_i}$ on each lattice site. $\vec{S_i}$ is a vector with n components S_{ia}, a=1,...,n. This is the so-called n-vector model. The length of each spin vector is fixed:

$$\sum_{a=1}^{n} S_{ia}^2 = n \qquad\qquad [4.70]$$

The spins interact with an external field $\vec{h}$ and with spins on connected lattice sites (nearest neighbors). The total energy of the whole system of spins $\{\vec{S_i}\}$ is then given by the Hamiltonian H:

$$H(\{\vec{S_i}\}) = -K\sum_{[ij]} \vec{S_i} \times \vec{S_j} - \sum_{i} \vec{h} \times \vec{S_i} \qquad\qquad [4.71]$$

K is a positive constant and [i,j] denotes nearest neighbor pairs. The thermodynamic behavior of this model follows from the partition function Z:

$$Z = \int d_{\vec{\Omega}} \exp\left(-\frac{H\{\vec{S_i}\}}{RT}\right) \qquad\qquad [4.72]$$

where $\int d_{\vec{\Omega}}$ represents integration over all possible orientations of all spins $\vec{S_i}$. The average over all possible spin orientations, with equal weights is denoted by $<>_0$. Thermal averages are indicated by $<>$ without subscript. The relationship between both averages, for any function $f(\{S_i\})$ is:

$$<f(\{\vec{S_i}\})> = \frac{<f(\{\vec{S_i}\}) \exp\left(-\dfrac{H(\{\vec{S_i}\})}{RT}\right)>_0}{<\exp\left(-\dfrac{H(\{\vec{S_i}\})}{RT}\right)>_0} \qquad\qquad [4.73]$$

The partition function Z of Eq. 4.72 can now be expressed as:

$$Z = \Omega <\exp\left(-\frac{H(\{\vec{S_i}\})}{RT}\right)>_0 \qquad\qquad [4.74]$$

where Ω is the total volume of the phase space of the spins and irrelevant for the succeeding calculations.

The exponential in Eq. (4.74) may be expanded, using Eq. (4.71) for the hamiltonian. This then leads to a mountain of terms containing averages such as:

$$<S_{ia}S_{jb}>_0 \quad <S_{ia}S_{jb}S_{kc}>_0 \qquad\qquad [4.75]$$

Since the averaging is over all possible orientations and each spin has an identical phase space, these averages may be replaced by similar averages over one particular spin, say S. The conclusion is that the partition function Z in Eq. (4.74) can be calculated if all averages of the type:

$$<S_a>_0 \quad <S_aS_b>_0 \quad <S_aS_bS_c>_0 \tag{4.76}$$

are known. A standard procedure to evaluate such spin averages is the introduction of the so-called characteristic function f(k), with $=(k_1,...,k_n)$:

$$f(k) = <\exp(i\vec{k}\times\vec{s})>_0 \tag{4.77}$$

The spin averages can be found from the coefficients of a Taylor expansion of f(k). Note that f is a function of the length k of k only, since the averaging over all orientations cannot leave an orientation dependence of f. So the problem of calculating Z is replaced by the problem of calculating f(k) which is, as yet, no fundamental improvement. One readily shows that the following differential equation for f(k) applies:

$$\frac{\partial^2 f}{\partial k^2} = \left(\frac{n-1}{k}\right)\frac{\partial f}{\partial k} + nf = 0 \tag{4.78}$$

with boundary conditions:

$$f(k=0) = 1 \quad \frac{\partial f}{\partial k}(k=0) = 0 \tag{4.79}$$

The important point here is that n lost its meaning as the number of vector components. Eq. (4.79) may be solved for any value of n and the solution for n = 0 turns out to be particularly simple:

$$f(k) = 1 - \frac{1}{2}k^2 \tag{4.80}$$

This result implies that all averages involving products of 3 and more spins must vanish in the n = 0 vector model.

Now return to the expression for Z, Eq. (4.74), with $\vec{h}= 0$ for a slightly different evaluation using the above results. First we note that:

$$\frac{Z}{\Omega} = <\exp\left(-\frac{H(\{\vec{S_i}\})}{RT}\right)>_0 = <\prod_{[ij]}\exp\left(\frac{K\vec{S_i}\times\vec{S_j}}{RT}\right)>_0 \tag{4.81}$$

The exponential is expanded and the vector product written out:

$$\frac{Z}{\Omega} = <\prod_{[ij]}\left[1 + \frac{K}{RT}\sum_a S_{ia}S_{ib} + \frac{1}{2}\left(\frac{K}{RT}\right)^2 \sum_{ab} S_{ia}S_{ib}S_{ja}S_{jb} + ...\right]>_0$$

Then the product over nearest neighbor terms [i,j] is expanded but now we know that only those terms will survive the subsequent averaging ($<>_0$) process in which every S_{ia} occurs exactly twice. The surviving terms are thus produced from non-intersecting closed loops of nearest neighbor bonds $[i_1 \ i_2] \ [i_2 \ i_3] \ [i_{k-1} \ i_k] \ [i_k \ i_1]$. Every such loop of length N generates by the averaging process a contribution of $(K/RT)^N$ for each of the n components. This implies that the partition function may be written as:

$$\frac{Z}{\Omega} = 1 + \sum_k \sum_{N_1...N_k} A(N_1,..,N_k)n\left(\frac{K}{RT}\right)^{N_1+...+N_k} \qquad [4.83]$$

where $A(N_1,...,N_k)$ is the number of ways to put k closed loops of $N_1,...,N_k$ bonds on the lattice without intersections. Now, the link with polymers is slowly becoming clear as lnA is the configurational entropy of k closed loop self-avoiding walks of specified lengths. However, for the n = 0 vector model, Eq. (4.83) becomes:

$$\frac{Z}{\Omega} = 1 \qquad [4.84]$$

which is not a very useful result. It is more interesting to consider spin-spin correlation functions:

$$<S_{ia}S_{ib}> = <S_{ia}S_{ib}\exp\left(-\frac{H}{RT}\right)>_0 \qquad [4.85]$$

This may be worked out for the n = 0 model in exactly the same way as Z was worked out. The consequence of the extra factors S_{ia} and S_{ib} that now appear is that the surviving terms correspond to self-avoiding walks from site i to site j (instead of closed loops). The following basic result is derived:

$$<S_{ia}S_{ib}>|_{n=0} = \sum_N A_N(i \rightarrow j)\left(\frac{K}{RT}\right)^N \qquad [4.86]$$

where $A_N(i \rightarrow j)$ is the total number of self-avoiding walks on the lattice of length N from site i to site j. Eq. (4.86) was developed by de Gennes.[41]

Note that A_N refers to a single chain on the lattice. A polymer solution or blend consists of a concentrated mixture of self-avoiding and mutually avoiding walks. By a trick, introduced by Des Cloizeaux,[43] the latter system can also be represented by a spin problem. If an external field $\vec{h}$ is introduced which is directed along one of the spin components one derives along the same ways as above:

$$\frac{Z(h)}{Z(0)} = 1 + \sum_{p\geq 1} \sum_{N\geq 0} \left(\frac{h}{RT}\right)^{2p}\left(\frac{K}{RT}\right)^{N} A(p,N) \qquad [4.87]$$

where $A(p,N)$ is the total number of ways to put p chains of all together N bonds on the lattice.

Eqs. (4.86÷4.87) are central results. They show that the problem of calculation of configurational entropies of self-avoiding chains on a lattice is mathematically equivalent to calculating spin-spin correlation functions on the same lattice. The latter problem, where the aspect of chain connectivity is lost, is much easier (but not necessarily easy !) to solve. Furthermore this type of spin models have already been rather intensively investigated and the above derivation allows translation of the results to polymer thermodynamic theory.

This approach has been heavily investigated by Freed and coworkers.[44-48] The basic exercise is to develop variations of the above lattice spin model that correspond to certain polymer configurational problems. For example, Eq. (4.87) describes a polydisperse system with uncontrolled chain length distribution. Freed noticed that it is possible to fix the distribution by using complex valued spins (2n components, but again in the limit n→0). Other features such as branching, specific monomer or solvent shapes[44,45] (all defined on the lattice) and even interactions[46] can be described by other variations of the spin model. A diagrammatic technique has been presented to facilitate the evaluations. Voids, giving the system compressibility, have also been introduced.[48] The quoted result for ΔS_M in Reference 48 appears to be incorrect. A detailed discussion of these complicated field theoretical analyses is beyond the scope of this book. Here we shall give the main results.

1. By adequate grouping of terms in the expansions, the configurational entropy can be expressed as a systematic expansion in the reciprocal lattice coordination number z^{-1}. The results are, in principle, exact. However, the analysis becomes extremely involved for higher order terms so one has to rely on the assumption that the first 2 or 3 terms form a sufficient approximation.

2. The zero order term corresponds to the mean field theory and reproduces the Flory-Huggins entropy of mixing.

3. For a polymer mixture of two polymers with respectively r_1 and r_2 segments (lattice sites) per chain, the correction to the Flory-Huggins entropy of mixing to first order in z^{-1} reads:

$$\Delta S_M^{ex} = -\frac{1}{z}\frac{(r_1-r_2)^2}{r_1^2 r_2^2}\Phi_1\Phi_2 \qquad [4.88]$$

For a polymer solution ($r_2 = 1$), Eq. (4.88) becomes:

$$\Delta S_M^{ex} = -\frac{1}{z}\left(1 - \frac{1}{r_1}\right)^2 \Phi_1 \Phi_2 \qquad [4.89]$$

It is interesting to note that this is exactly the Huggins correction, Equation 3.40, which was derived 50 years ago as a rather ad hoc correction to the mean field result.

4.6 PRISM THEORY

Another alternative to modelling the thermodynamic behavior of polymeric mixtures by means of lattice models is to start from an atomistic level. This is done in the Polymer Reference Interaction Site Model (PRISM), a very recent approach to the thermodynamics of polymer mixtures, which in principle extends beyond the limitations of more traditional lattice models.

PRISM theory, developed over the past years by Schweizer and Curro,[49-58] is an off-lattice continuum theory of polymeric liquids, both as pure phase and as mixtures. It is based on the small molecule RISM theory developed by Chandler and Andersen.[59,60] In brief, given a knowledge of the intramolecular structure of a single chain, PRISM describes the packing of either like or unlike chains in the melt and the corresponding thermodynamic behavior.

PRISM theory is based on the statistical mechanical theories of liquids, the basis of which provide enough material to fill an entire book.[61] The basis of such theory is centered around the interatomic pair distribution function $g(r,r')$. It specifies the relative probability of finding an atom at position r, given that another atom is at position r'. In an isotropic liquid, $g(r, r')$ is a function of the magnitude $r - r'$ only; that is, it has no angular dependence or dependence on the absolute location in the liquid, and is generally written as $g(r)$. At short distances, due to the hard core repulsions between pairs of atoms, the probability that chains overlap is zero, and $g(r)$ vanishes. At intermediate distances, $g(r)$ has a roughly oscillatory shape corresponding to the packing of layers immediately surrounding an atom in a liquid. At long distances, all correlations between atomic positions are lost, so that $g(r)$ approaches unity.

All modern liquid state theories are based on the Ornstein-Zernike (OZ) equation:

$$h(r) = c(r) + \rho \int c(r')h(r - r')dr' \qquad [4.90]$$

In this equation, ρ is the number density of particles, $h(r)$ is defined as $g(r) - 1$, and $c(r)$ is the direct correlation function. By itself, this equation is exact but trivial in that $c(r)$ is defined to be the function that makes the OZ equation true. To give the equation more content, a closure relation must be used. The OZ equation may be expressed more simply as follows:

$$h(r) = c(r) + c(r)*h(r) + c(r)*c(r) + c(r)*c(r)*c(r) + \ldots \qquad [4.91]$$

where the asterisks denote convolution integrals as in Eq. (4.90). The first convolution on the right hand side of Eq. (4.91) represents the direct correlation between two atoms, without interference of other atoms. The second is the first indirect correlation, in which two atoms interact via a third atom. The contribution of higher order indirect correlations (hopefully) vanishes with increasing number of atoms. The OZ equations contains two unknown functions: $c(r)$ and $h(r)$. To solve it requires another equation, a closure, relating $h(r)$ and $c(r)$. A closure is an approximation that allows the OZ equation to be solved. Several closures exist,[61] but the most common one is the one known as the Mean Spherical Approximation (MSA):

$$h(r) = -1 \quad r < \sigma \qquad\qquad\qquad [4.92]$$

$$c(r) = \frac{u(r)}{k_B T} \quad r > \sigma$$

where σ is the hard core diameter of the atoms and $u(r)$ is the potential energy outside the hard core.

The structure made up of rigid molecules is more complicated than that of atomic liquids, due to the loss of spherical symmetry when two or more atoms combine to form a molecule. This loss of symmetry implies that the pair correlation function takes on an angular distribution, i.e., $g(r)$ becomes $g(r,\Omega,r',\Omega')$, where Ω and Ω' are the orientational coordinates of the molecules whose centers of mass lie at r and r', respectively. Explicit treatment of this angular dependence can be extremely difficult.

An alternative to explicit treatment of molecular orientations is to regard each molecule in a liquid as consisting of a number of sites, which may be, but do not have to be, atomic centers. This simplification comes from the assumption that interactions between sites on different molecules are spherically symmetric in form. The liquid structure is than described, not by means of a complicated orientation dependent molecular pair distribution function, but by a set of site-site pair distribution functions $g_{\alpha\beta}(r)$, where the subscripts α and β label the type of site. This general formalism is known as Interaction Site Model (ISM). Although the number of sites in a molecule may correspond to the number of atoms, this is not essential. Depending on the nature of the molecule, it is often useful to have fewer sites than atoms. For instance, the CH_2 group in the polyethylene chain could be treated as one site. The model does than depend upon availability of suitable united atom potentials. An ISM representation should capture enough of the molecule's structure and represent its interactions with other molecules sufficiently well for the application of interest.

An ISM is called a Reference ISM, or RISM, when the sites are treated as hard spheres, possibly with the hard sphere diameters as adjustable parameters. Here the term, "reference" indicates that the hard sphere system may form a mathematical reference system for a relevant perturbation expansion. Chandler and Andersen,[59,60] formulated an approximate RISM integral equation theory for RISM molecules. It is based on a molecular generalization of the OZ equation and on an MSA-like closure relation. This RISM theory has proven very successful in calculating the structures of molecular liquids. In its overall structure, RISM theory is the same for both small molecule and polymeric liquids. The details of the Polymer Reference Interaction Site Model are discussed below.

PRISM is somewhat more difficult than RISM because both intra and inter site-site pair correlation functions need to be treated. The generalized OZ equation of PRISM is:

$$\mathbf{H}(r) = \int dr' \int dr'' \Omega(r - r')\mathbf{C}(r' - r'')[\Omega(r'') + \mathbf{H}(r'')] \qquad [4.93]$$

where $\mathbf{H(r)}$ is the matrix of intermolecular site-site pair correlation functions for all possible pairs of site types, $\mathbf{C(r)}$ is the corresponding matrix of direct pair correlation functions, and $\Omega\mathbf{(r)}$ is the matrix of intramolecular site-site distribution functions. The principle difference between Eq. (4.93), for polymers, and Eq. (4.90), for molecules, is that the complete chain of direct and indirect correlations is convoluted with the intramolecular site-site distribution function. The physical picture behind this is that the intramolecular arrangement has an effect on the intermolecular arrangements and *vice versa*. In other words, the chain stiffness and persistence determines the local structure of surrounding polymers, just as unlike polymer site-site interactions determine the actual arrangement of the individual chain: it is a matter of competition. In this way, the complete interaction balance, hence the local structure and free energy, can be calculated on a molecular scale. Precise knowledge of site-site interaction potentials is a prerequisite here.

Because the elements in the generalized OZ Eq. (4.93) are matrices, this "equation" is really a set of coupled equations. The number of distinct equations is m(m+1)/2, where m is the rank of the matrices. This rank is equal to the number of distinct site types in the system. In this context, two sites on a chain are considered distinct if their direct correlation functions with sites on other chains differ from each other. That is, sites j and k are distinct if $C_{jl}(r)$ and $C_{kl}(r)$ differ, where l is the index of a third site. Two sites are equivalent if these direct correlation functions (for all l) are the same. For example, in a monodisperse melt of simple ring polymers, all sites are equivalent; m is equal to 1, and the generalized OZ equation reduces to only a single integral equation, which is easily solvable.

In a melt of linear simple chains, the situation is in principle more complex in that sites near the chain ends are chemically different from those in the chain interior. Treatment of such end-effects would render the OZ equation unsolvable for practical purposes. Curro and Schweizer make the approximation to neglect end effects, and to treat all sites equivalent. This makes RISM theory tractable when applied to long linear polymers. Together with the closure relation, Eq. (4.92), now applied to the relevant site-site interactions, the set of PRISM equations can be solved. However, a useful description of the structure of polymeric liquids and hence, the thermodynamic behavior of mixtures, is only obtained when reliable site-site intra- and interatomic interaction potentials are available.

A sizable amount of effort has been put in the prediction of the local structure, and correlated properties, of simple polyethylene.[62-64] Predictions can be compared with Molecular Dynamics simulations, in which the actual potentials used are known. Most critical is a correct description of the intramolecular structure factor $\Omega(r)$. At a large enough length scale, all chains behave gaussian, because there is no correlation anymore between sites (unless the chain is infinitely stiff). On shorter length scales however, actual chain stiffness, or semi-flexibility, plays definitely a role, so that the relatively short range local structure around segments is in principle different from the overall structure of the liquid. Hence, in order to describe the actual structure of already simple polymers like polyethylene correctly, such semi-flexibility has to be taken into account. Possible methods here are for example interpolation between the $\Omega(r)$ of rigid-rod chains and gaussian chains,[65] or to employ more sophisticated methods in which a limited number of torsional rotations is allowed (such as trans-gauche conformations). The former method leads to an adjustable parameter, namely the persistence length or flexibility parameter, that needs to be fitted to data points, the latter method is in principle more exact, but very time-consuming. By employing such techniques, PRISM has been successful in predicting the structure of polyethylene melts, when compared to molecular dynamics simulations,[66] see Figure 4.8. Melting points of PE and PTFE were predicted successfully as well.[63]

Given sufficient knowledge about site-site potentials, one is in the position to calculate various properties of melts and mixtures. In

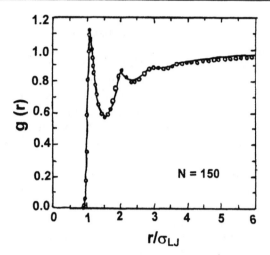

Figure 4.8. PRISM prediction of the radial distribution function of a polyethylene chain with 150 units. Points are MD results[66]. (Adapted, by permission, from K. G. Honnell, J. G. Curro and K. S. Schweizer, *Macromolecules*, **23**, 3496 (1990)).

scattering experiments, one measures the structure factor S(k) of a liquid, where k is the wavevector (see also Chapter 6). In the context of PRISM theory, given the intra- and intermolecular site-site correlation functions, S(k) is defined by:

$$S(k) = \Omega(k) + H(k) \qquad\qquad [4.94]$$

where

$$H(k) = \int dr \, \exp(-ikr)H(r) \qquad\qquad [4.95]$$

with similar expressions for $\Omega(k)$ and S(k). The attractive aspect to work via the structure factor S(k) is that, just like for the partition function Z, if this function is known, everything about the liquid (be it pure component or mixture) is known as well. For instance, the Isothermal Compressibility κ_T, is defined as:

$$\kappa_T = -\frac{1}{V}\left(\frac{\partial V}{\partial P}\right)_T \qquad\qquad [4.96]$$

which characterizes, o.a., the dependence of the free energy on volume and hence, the EoS behavior of the liquid. Thermodynamically, S(k) and κ_T are related:[67]

$$S(k) = RT\rho\kappa_T \qquad\qquad [4.97]$$

$$\lim k \to 0$$

Hence, extrapolation of $S(k)$ to $k=0$ gives the isothermal compressibility, or equivalently, the EoS behavior, of the liquid. Along these lines, the EoS behavior of alkanes and polyethylene could be correctly predicted.[68,69]

The simplest approach to liquid-liquid phase separation is to identify it with the point of mechanical instability. This corresponds to the spinodal point at which the scattering intensity diverges, or equivalently, where $S(k)$ becomes infinite at small k values. Although, such calculations are not straightforward and the translation from PRISM theory to actual thermodynamics of mixtures is not unambiguous, exact results can be obtained for certain special cases. The "symmetric blend" (equal chain lengths N and simple dispersive interactions) is such a case. A striking result from PRISM theory [57] is that the UCST of such blend is found to be proportional to $\sqrt{N}$ (or critical χ parameter, χ_c proportional to $1/\sqrt{N}$) rather than the linear N dependence predicted by FH theory, see Eq. (4.3). This result is obtained for weakly asymmetric blends as well.

Apparently, the dependence of χ_c on N is reduced by a factor $\sqrt{N}$, a factor proportional to the radius of gyration R_g. In close analogy with the correlation hole effect described by de Gennes,[40] Curro and Schweizer suggest that the effect is caused by spatial correlations between segments within an R_g distance. Consequently, the effective interaction parameter becomes a function of R_g.

It is hard to judge in how far PRISM theory will prove useful in the more daily practice of predicting phase behavior of realistic polymer mixtures. The full implications of PRISM theory and its range of applicability are simply not known at this time, nor is the extent to which its predictions are quantitatively reliable in general. This is to a certain extent due to the general lack of reliable united atom potentials and the difficulty of routinely calculating the intramolacular structure of polymer chains. Which brings us, indirectly, back to the "good-old" lattice model: pretty smart after all?

REFERENCES

1. M. L. Huggins, *J. Chem. Phys.*, **9**, 440 (1941).
2. M. L. Huggins, *Ann. N. Y. Acad. Sci.*, **43**, 1 (1942).
3. P. J. Flory, *J. Chem. Phys.*, **9**, 660 (1941).
4. P. J. Flory, *J. Chem. Phys.*, **10**, 51 (1942).
5. R. Koningsveld, L. A. Kleintjes, and H. M. Schoffeleers, *Pure Appl. Chem.*, **39**, 1 (1974).
6. I. Hashizume, A. Teramoto, and H. Fujita, *J. Polym. Sci., Polym. Phys. Ed.*, **19**, 1405 (1981).
7. Th. G. Scholte, *J. Polym. Sci.*, **9**, 1553 (1971).
8. K. W. Derham, J. Goldsbrough, and M. Gordon, *Pure&Appl. Chem.*, **38**, 97 (1974).
9. L. A. Kleintjes and R. Koningsveld, *J. Electrochem. Soc.*, **121**, 2353 (1980).

10. L. A. Kleintjes and R. Koningsveld, *J. Colloid. Pol. Sci*, **258**, 711 (1980).
11. R. van der Haegen, L. A. Kleintjes, L. van Opstal, and R. Koningsveld, *Pure. Appl. Chem.*, **61**, 159 (1989).
12. A. Bondi, *J. Phys. Chem.*, **68**, 441 (1964).
13. R. Koningsveld, H. A. G. Chermin and M. Gordon, *Proc. Royal Soc.*, **A319**, 331 (1970).
14. P. J. Flory, *Principles of Polymer Chemistry*, Cornell University Press, London 1953.
15. I. Prigogine, A. Bellemans and V. Mathot, *The Molecular Theory of Solutions*, North Holland Publishing, Amsterdam, 1957.
16. P. J. Flory, R. A. Orwoll, and A. Vrij, *J. Am. Chem. Soc.*, **86**, 3567 (1964).
17. P .J. Flory, *J. Am. Chem. Soc.*, **87**, 1833 (1965).
18. B. E. Eichinger and P. J. Flory, *Trans. Faraday Soc.*, **65**, 2035 (1968).
19. P. J. Flory, *Disc. Faraday Soc.*, **49**, 7 (1970).
20. L. P. McMaster, *Macromolecules*, **6**, 760 (1973).
21. D. Patterson and A. Robard, *Macromolecules*, **11**, 690 (1978).
22. G. T. Dee and D. J. Walsh, *Macromolecules*, **21**, 811 (1988).
23. G. T. Dee and D. J. Walsh, *Macromolecules*, **21**, 815 (1988).
24. I. C. Sanchez and R. H. Lacombe, *J. Phys. Chem*, **80**, 2352 (1980).
25. I. C. Sanchez and R. H. Lacombe, *J. Polym. Sci., Polym. Lett. Ed.*, **15**, 17 (1976).
26. R. H. Lacombe and I. C. Sanchez, *J. Phys. Chem.*, **80**, 2568 (1976).
27. I. C. Sanchez and R. H. Lacombe, *Macromolecules*, **6**, 1145 (1978).
28. I. C. Sanchez in *Polymer Blends*, Vol. 1, D. R. Paul and S. Newman, Eds., Academic, NY 1978.
29. T. Somcynski and R. Simha, *J. Appl. Phys.*, **42**, 4545 (1971).
30. R. Simha and T. Somcynski, *Macromolecules*, **2**, 341 (1969).
31. A. Stroeks and E. Nies, *Polym. Eng. Sci.*, **28**, 1347 (1988).
32. E. Nies and A. Stroeks, *Macromolecules*, **23**, 4088 (1990).
33. A. Stroeks and E. Nies, *Macromolecules*, **23**, 4092 (1990).
34. W. B. Jensen, *The Lewis Acid-Base Concepts*, Wiley, NY, 1980.
35. E. A. Guggenheim, *Proc. R. Soc. London A*, **183**, 213 (1944).
36. G. Ten Brinke and F. E. Karasz, *Macromolecules*, **17**, 815 (1984).
37. A. Wakker and M. A. van Dijk, *Polym. Networks&Blends*, **2**, 123 (1992).
38. M. M. Coleman, J. F. Graf, and P. C. Painter, *Specific Interactions and the Miscibility of Polymer Blends*, Technomic Publishing, 1991, and references therein.
39. I. C. Sanchez and A. Balasz, *Macromolecules*, **22**, 2325 (1989).
40. P. G. de Gennes, *Scaling Concepts in Polymer Physics*, Cornell University Press, Ithaca, 1979.
41. P. G. de Gennes, *Phys. Lett.*, **A38** 339 (1972).
42. K. F. Freed, *J. Phys.* **A18** 871 (1985).
43. J. des Cloizeaux, *J. Physique (Paris)* **36** 281 (1975).
44. M. G. Bawendi and K. F. Freed, *J. Chem. Phys.*, **85**, 3007 (1986).
45. M. G. Bawendi and K. F. Freed, *J. Chem. Phys.*, **86**, 3720 (1987).
46. M. G. Bawendi and K. F. Freed, *J. Chem. Phys.*, **87**, 5534 (1987).
47. M. G. Bawendi and K. F. Freed, U. Mohanty, *J. Chem. Phys.*, **84**, 7036 (1986).
48. M. G. Bawendi and K. F. Freed, *J. Chem. Phys.*, **88**, 2741 (1988).
49. K. S. Schweizer and J. G. Curro, *Phys. Rev. Lett.*, **58**, 256 (1987).
50. J. G. Curro and K. S. Schweizer, *Macromolecules*, **20**, 1928 (1987).
51. J. G. Curro and K. S. Schweizer, *J. Chem. Phys.*, **87**, 1842 (1987).
52. K. S. Schweizer and J. G. Curro, *Macromolecules*, **21**, 3070 (1988).
53. K. S. Schweizer and J. G. Curro, *Macromolecules*, **21**, 3082 (1988).
54. K. S. Schweizer and J. G. Curro, *J. Chem. Phys.*, **89**, 3342 and 3350 (1988).

55. K. S. Schweizer and J. G. Curro, *Phys. Rev. Lett.*, **60**, 809 (1988).
56. J. G. Curro and K. S. Schweizer, *J. Chem. Phys.*, **88**, 7242 (1988).
57. K. S. Schweizer and J. G. Curro, *J. Chem. Phys.*, **89**, 5059 (1989).
58. J. G. Curro and K. S. Schweizer, *Macromolecules*, **23**, 1402 (1990).
59. D. Chandler and H. C. Andersen, *J. Chem. Phys.*, **57**, 1930 (1972).
60. D. Chandler, *J. Chem. Phys.*, **59**, 2742 (1973).
61. J. P. Hansen and I. R. McDonald, *Theory of Simple Liquids*, Academic, London, 1976.
62. K. G. Honnell, J. G. Curro and K. S. Schweizer, *Macromolecules*, **23**, 3496 (1990).
63. J. D. McCoy, K. G. Honnell, K. S. Schweizer, and J. G. Curro, *J. Chem. Phys.*, **95**, 9348 (1991).
64. J. D. McCoy, K. G. Honnell, J. G. Curro, and K. S. Schweizer, *Macromolecules*, **25**, 4905 (1993).
65. R. Koyama, *J. Phys. Soc. Jpn*, **34**, 1029 (1973).
66. K. Kremer and G. S. Crest, *J. Chem. Phys.*, **92**, 5057 (1990).
67. L. P. Landau and E. M. Lifshitz, *Statistical Physics*, Pergamon, NY 1980.
68. J. G. Curro, A. Yethiraj, K. S. Schweizer, J. P. McCoy, and K. G. Honnell, *Macromolecules*, **26**, 2655 (1993).
69. A. Yethiraj, J. G. Curro, K. S. Schweizer, and J. P. McCoy, *J. Chem. Phys.*, **98**, 1635 (1993).

Chapter 5

COMPUTER SIMULATIONS

5.1 INTRODUCTION

The traditional way of progress in modern science has been through the interplay between experimental results and analytical theories. A physical model of the system was translated into a set of mathematical equations and from then on the solution of the problem was basically a matter of mathematical skills. The development of new theories and better models was restricted by what could mathematically be solved. Originally the mathematics were done with pencil and paper and cunning manipulation of equations. The advent of computers gave rise to a very much increased capability to solve complex equations by numerical algorithms. In principle, any problem that can be translated into a set of mathematical equations can be solved numerically. This brought the translation of nature in a model and a model into a set of equations back into focus. Particularly for those systems in nature that are so complex that a translation into a finite set of equations involves large approximations.

The development of fast computers gave rise to a whole new area of science: computer simulations. The idea now is to translate the complex system of nature into a simplified, well-defined but still complex model system. The behavior of this model system is then simulated with a computer program. From the vast amount of data generated by the simulations, several, usually averaged, properties are extracted. These properties mimic the corresponding experimentally measurable properties of the original system in nature. Experiments test the validity of our simplified (solvable) models and the approximations involved. Computer simulations allow us to go one step further and to test the validity of the complex (unsolvable) model from which the simplified model was derived by comparing experimental and simulated properties. Computer simulations also allow testing the validity of the approximations involved in deriving the simplified model by comparing simulated and calculated properties.[1-4] This is illustrated in Figure 5.1.

As an example, consider a polymeric melt. The melt is pictured as a mixture of chains on a 3 dimensional cubic lattice. This is already a large approximation but still constitutes a very complex model. By

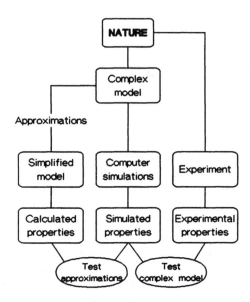

Figure 5.1. Relations between theoretical models, computer simulations, and experiment.

making suitable further approximations (such as treating the chains as ideal random walks), we may derive a model which can be solved and from which properties (e.g., diffusion constants, densities) may be calculated. The lattice model of the polymer melt can also be studied with computer simulations. In this way one finds the properties of the lattice model. By comparing experimental data with simulated and calculated values it is possible to assess which part of the deviations is due to the use of a lattice model and which part is due to the approximations involved in deriving the simplified model.

Another possible use of computer simulations is in what one could call hybrid theories. These are analytical theories that use some quantities that can only be calculated by means of computer simulations (e.g., the number of self-avoiding walks of length 50). This quantity plays the role of an empirical parameter in the final theoretical expressions.

This chapter has been arranged as follows. First some basic statistical mechanics results will be discussed. Statistical mechanics play a crucial role in the interpretation of computer simulations. Then the two most important types of computer simulations: molecular dynamics and Monte Carlo simulations are discussed. Finally, in the Section 5.6, the simulation of the thermodynamic behavior of polymeric systems is surveyed.

5.2 COMPUTER SIMULATIONS
AND STATISTICAL MECHANICS

5.2.1 INTRODUCTION

Statistical mechanics[5,6] provides tools to relate detailed information on the microscopic states of a system to macroscopic properties. Computer simulations can be used to generate the microscopic data.

The thermodynamic state of a system is defined by a small number of quantities such as the number of particles N, pressure P, and temperature T. The microscopic state of the system is defined by a very large number of quantities; for example the 6N numbers that describe position and momentum of each particle. These 6N numbers form a 6N dimensional space where each microscopic state corresponds to one point X in this so-called phase space.

The instantaneous value of some property C is a function of the phase space coordinate X: C(X). As the system evolves in time, X changes and hence also the value of P. If C(X) is known (from statistical mechanics) and X(t) is obtained from a computer simulation of the system over a certain time interval) then the time averaged value <C> can be calculated.

A practical system (N ~ 10^{23}) will always be much larger than the relatively small system that one is able to simulate (N ~ $10^{3.5}$). In a sense one could say that the practical system consists of a very large number of these small systems where each sub-system may be on a different phase space coordinate. An average property of the large system can therefore be seen as the average of the property over a large number of sub-systems. A collection of sub-systems is called an ensemble. The collection of points that form the ensemble wanders through phase space with time. If one observes the trajectory of an ensemble long enough one will see that some points are visited more often than other points. The points are distributed according to a probability density P(X). This function is determined by the choice of fixed thermodynamic parameters (e.g., N,V,T or N,P,T). This procedure of following the time evolution of one particular ensemble to obtain P(X) is only valid if all points in phase space that are compatible with the thermodynamic (N,P,T,V) constraints will indeed be visited by the wandering ensemble. Such a system is called ergodic. It is very difficult to prove that a system is ergodic but one generally believes that most systems are. Non-ergodic systems may be constructed but they are usually pathological.

If the ergodic condition is fulfilled, one could also construct the phase space probability density by considering a large number of (randomly chosen) sub-systems as suggested by Gibbs. One may then replace time averages by ensemble averages which is often more convenient from a computational point of view.

The purpose of computer simulations is to generate a representative number of phase space ensembles so that the probability distribution P(X) can be determined. The usual strategy is to generate an initial state of the system which is hopefully not too far from equilibrium. The system is then allowed to evolve over a succession of states. In molecular dynamics simulations, this succession of states is generated by the true equations of motion and creates a true time trajectory in phase space. Other techniques, such as Monte Carlo simulations use entirely different recipes which may not have any physical interpretation at all but are compatible with the thermodynamic constraints. Whatever the algorithm, there are always two (related) questions to be asked: 1. what is the effect of the choice of initial conditions on the results and 2. has a sufficiently large and representative part of phase space been explored in the limited number of steps of the simulation. It is extremely difficult to prove that this is the case. One generally creates confidence in the results by doing several simulations with different initial conditions and showing that the relevant properties are constant.

As discussed above, the phase space probability density P(X) depends on the choice of fixed thermodynamic parameters. The most used choices are:
1. fixed N,V, and E (the so-called microcanonical ensemble)
2. fixed N,V, and T (the canonical ensemble)
3. fixed N,P, and T (the isobaric isothermal ensemble) and
4. fixed chemical potential μ, V, and T (the grand canonical ensemble)

Before discussing these ensembles and their relation to thermodynamic quantities in some more detail we will first derive the so-called Boltzmann distribution law. First, because it is an important result and second because it gives an idea of the sort of mathematics that is involved in the derivation of the subsequent results which we shall give without proof.

5.2.2 THE BOLTZMANN DISTRIBUTION

Consider a system of N non-interacting particles in contact with a heat bath at temperature T. The volume V is fixed. In fact, this corresponds to the canonical ensemble of an ideal gas. The temperature T defines the total energy E of the system. Each particle has energy levels E_1, E_2,... etc. As the particles are non-interacting, these energy levels are unaffected by the presence of other particles. Let the particles be distributed over these energy levels, such that N_i particles are at energy level E_i. The notion of energy levels is natural to quantum mechanics, where the volume V would typically determine the spacing of the energy levels. In classical mechanics, there are no discrete energy levels. One may however set up a similar reasoning by subdividing the energy scale in (infinitesimally small) energy band dE_i. For clarity, we will consider discrete energy levels here. Constant number of particles N and energy E implies:

$$\sum_i N_i = N \quad \sum_i N_i E_i = E \qquad [5.1]$$

Since everything has been conveniently discretized now, we may use combinatorial algebra to calculate the number of ways in which the N particles can be divided over the energy levels. A particular distribution with N_i particles at energy level E_i can be realized in W_x different ways:

$$W_x = \frac{N!}{N_1! N_2! ... N_i! ...} = \frac{N!}{\prod_i N_i!} \qquad [5.2]$$

The most probable distribution (assuming equal weight for each state) corresponds to the maximum of W_x under the constraints of Eq. (5.1). This maximum can be worked out, using Lagrange multipliers.[6] The result is the Boltzmann distribution law:

$$\frac{N_i}{N} = \frac{e^{-\frac{E_i}{kT}}}{q} \qquad [5.3]$$

where q, given by:

$$q = \sum_i e^{-\frac{E_i}{kT}} \qquad [5.4]$$

is called the particle partition function. The temperature originally entered the derivation as an undefined constant which could be associated with temperature by comparison of the computed average energy with the average kinetic energy of an ideal gas.

It can be shown[6] that the probability distribution around this most probable distribution is extremely small so that the average and most probable distributions may be considered identical.

The Boltzmann distribution describes how the particles will divide among the energy levels if the system is left alone and the particles are able to access all available levels. The distribution is the result of the trade off between maximum entropy, which would imply uniform distribution over the energy levels and the requirement of constant energy.

The particle partition function q plays an important role. Knowledge of q(N,T,E) enables calculation of all relevant thermodynamic properties. For example, the average energy per particle <E> is given by:

$$<E> = \frac{\sum N_j E_j}{\sum N_j} = kT^2 \left(\frac{\partial \ln q}{\partial T} \right)_v \qquad [5.5]$$

This example also shows that the absolute value of q is irrelevant. The thermodynamics of the system is determined by how q depends on the thermodynamic parameters.

5.2.3 ENSEMBLES

The microcanonical ensemble is the set of systems with fixed number of particles N, volume V, and total energy E. The principle of equal a priori probabilities implies that each realization of the system is equally probable. This in turn implies that the probability of finding one particular realization X is given by:

$$P(X) = \frac{1}{Q_{NVE}} \qquad [5.6]$$

where Q_{NVE} is the so-called partition function of the microcanonical ensemble. Q_{NVE} is the total number of possible states of the system with total energy H(X) = E. H is the hamiltonian of the system and includes kinetic as well as potential energies. The logarithm of Q is proportional to the entropy S of the system:

$$S = k \ln Q_{NVE} \qquad [5.7]$$

The canonical ensemble is the set of systems with fixed N, V, and temperature T and is more closely related to experimental systems. It can be regarded as a large number of subsystems which are connected by walls that allow passage of (thermal) energy but not of particles. The whole (infinite) system has a constant energy E. The value of E determines the temperature T. All sub systems are identical and thus have the same set of allowed values of total energy E_i. The problem of calculating the most probable realization of the system is mathematically similar to that treated in the derivation of the Boltzmann distribution. Hence, the probability of finding a realization X_i with total energy $H(X_i)$ is given by:

$$P(X_i) = \frac{e^{-\frac{H(X_i)}{kT}}}{Q_{NVT}} \qquad [5.8]$$

where Q_{NVT} is the partition function of the canonical ensemble:

$$Q_{NVT} = \sum_{X_i} e^{-\frac{H(X_i)}{kT}} \qquad [5.9]$$

The corresponding thermodynamic function is the Helmholtz free energy F:

$$F = -kT \ln Q_{NVT} \qquad [5.10]$$

In the isothermal-isobaric ensemble, the volume V is not fixed and enters the list of quantities (in addition to the momenta and positions of the particles) describing the state of the system. The partition function Q_{NPT} is now given by:

$$Q_{NPT} = \sum_{X} \sum_{V} e^{-(H+PV)/kT} = \sum_{V} e^{-PV/kT} Q_{NVT} \qquad [5.11]$$

The corresponding thermodynamic function is the Gibbs free energy G:

$$G = kT \ln Q_{NPT} \qquad [5.12]$$

Finally, in the grand canonical ensemble the number of particles N is variable. The grand canonical partition function Q_{VT} is given by:

$$Q_{\mu VT} = \sum_{X} \sum_{N} e^{-(H-\mu N)/kT} = \sum_{N} e^{\mu N/kT} Q_{NVT} \qquad [5.13]$$

and the corresponding thermodynamic function is given by:

$$\frac{PV}{kT} = \ln Q_{\mu VT} \qquad [5.14]$$

The summations in the above equations are assumed to take due note of the indistinguishability of the particles: n_1 particles on position A, n_2 particles on position B on the one hand and n_2 particles on position A, n_1 particles on position B on the other, are considered identical states and are counted only once. The summations also suggest that the amount of possible states is finite. According to quantum mechanics, the system has discrete energy levels and the allowed states are countable. In the quasi classical expression for Q, the summation is replaced by an integral over all 6N positions r and momenta p:

$$Q = \sum_X = \frac{1}{N!h^{3N}} \int dr dp \qquad\qquad [5.15]$$

The integration is over all 6N coordinates, the indistinguishability of the N particles is explicitly taken into account by the factor N!, which is the number of (indistinguishable) ways to distribute N particles over N coordinates. The somewhat mysterious presence of Planck's constant h is to allow a correct treatment of the ideal gas, which serves as a calibration point between classical and quantum mechanics.

The total energy H(X) of a classical system can always be expressed as a sum of kinetic (p-dependent) and potential (r-dependent) contributions. This allows the partition function integrals of Q_{NVT} and Q_{NPT} to be written as a product of kinetic (K) and potential (U_p) terms. For example:

$$Q_{NVT} = \frac{1}{N!} \frac{1}{h^{3N}} \int dp \; e^{-\frac{K}{kT}} \int dr \; e^{-U_p/kT} \qquad\qquad [5.16]$$

For a system without interactions ($U_p = 0$) one obtains:

$$Q_{NVT} = \frac{V^N}{N!h^{3N}} \int dp \; e^{-K/kT} \qquad\qquad [5.17]$$

On the other hand, such a system is by definition an ideal gas. The partition function Q_{NVT}^{id} of an ideal gas can be obtained from quantum mechanics:

$$Q_{NVT}^{id} = \left(\frac{2\pi m kT}{h^2}\right)^{3N/2} \frac{V^N}{N!} \qquad\qquad [5.18]$$

Indeed, with $K = (p_1^2 + p_2^2 + p_3^2)/2m$ per particle with mass m, Eq. (5.17) gives the ideal gas result Eq. (2.18).

The partition function may thus be written as the product of an ideal gas contribution Q_{NVT}^{id} and an excess part Q_{NVT}^{ex}:

$$Q_{NVT} = Q_{NVT}^{id} + Q_{NVT}^{ex} \qquad\qquad [5.19]$$

with:

$$Q_{NVT}^{ex} = \frac{1}{V^N} \int dr \; e^{-U_p/kT} \qquad\qquad [5.20]$$

Instead of Q^{ex}, one often uses the so-called configurational integral Z_{NVT}:

$$Z_{NVT} = \int dr \; e^{-V(r)/kT} \qquad\qquad [5.21]$$

5.3 MOLECULAR DYNAMICS

Molecular dynamics (MD) finds its inspiration in the spirit of Laplace who stated that if shape, position, and velocity of all atoms of a system were known, the development of the system with time can be calculated. From the full description of the behavior of the system, macroscopic properties, such as density, cohesive energy, and specific heat may be extracted. A molecular dynamics simulation thus provides the link between a microscopic picture of the system (atoms and force fields) and its macroscopic physical properties.

Unfortunately, full scale molecular dynamics simulations of polymeric materials are still far beyond the horizon of computational possibilities. To illustrate this, consider a polyethylene chain with a very modest molecular weight of 60 kg/mole. Such a chain has an unperturbed radius of gyration of about 10 nm. For a reasonable simulation we would at least need a volume of 10^3 nm^3. Such a volume element contains about 40,000 CH_2 groups and hence about N = 120,000 atoms. This tiny little volume element thus contains about 10^{10} (N(N – 1)/2) atom-atom pair interactions. So even if we me assume that the effects of the electrons are averaged out (Born-Oppenheimer approximation) and that only pair interactions contribute to the net force on any atom, we would need to calculate 10^{10} force components per time step.

This brings us to the next complication: the time scale. Our polymer chain of about 4,300 segments has a reptation diffusion coefficient which is of the order[7] of 50 nm^2/s. This means that the polymer will take about 2 seconds to reptate over a distance of its own size (10 nm). On the other hand, a polymer liquid is a dense system and one needs to repeat the force calculation after a very short period of time. An indication of the maximum time integration step can be obtained from typical phonon vibration frequencies which are of the order[8] of 10^{12} Hz. We would thus need time steps less than 10^{-13} s.

So, in order to simulate a system with the size of a 60 kg/mol PE coil (10 nm) over a period of time which is relevant for macroscopic relaxation processes (1 sec) one would need to calculate about 10^{23} interactions. A hyper computer with a rating of 100 GFlops (10^{11} floating point operations per second) would need hundreds of centuries to do this.

So much for full scale molecular dynamics of polymeric systems. It is clear that polymers, being large and slow are the most unsuitable molecules for MD simulations on earth. For MD simulations to become feasible, one or more additional approximations must be made. Examples are: use smaller chains (N of the order of 10^2), short range interactions (e.g., only nearest neighbors) and lumped groups (e.g., treat -CH_2- as one particle). Furthermore, some problems do not need

simulations over macroscopic time intervals. Examples are: the diffusion of a small molecule through a polymer, local chain dynamics and
melting and crystallization behavior.

For polymer thermodynamic problems, MD simulations are
rarely used. Again, because of the relatively small part of phase space
that is sampled during a simulation.

5.4 MONTE CARLO SIMULATIONS

Monte Carlo methods provide a way to estimate partition functions
without having to perform the full integration. This may be illustrated
by the classical example of estimating π.

The area of one quadrant of a circle with radius 1 is $\pi/4$. Which
can be calculated by numerical integration:

$$\pi/4 = \int_0^1 (1 - x^2)^{1/2} dx \qquad [5.22]$$

using Simpson's rule. Alternatively, one could generate a large number of randomly chosen values x_i and y_i between 0 and 1 and calculate
the fraction of values for which $(1 - x_i^2)^{1/2} < y_i$. This is the fraction of
points that fall under the circle, see Figure 5.2.

The power of the latter method is that after a relatively small
number of function evaluations one already has a crude estimate of
π while brute force integration requires sequential evaluation of the
entire function. For a simple one dimensional integration as in Eq.
(5.1) this is not a problem but for the multi-dimensional partition
function integrals brute force integration is unfeasible. In such a case,
the Monte Carlo method is a very useful approach.

The configurational partition function of the canonical ensemble
Z_{NVT} could be estimated by generating a large number K of randomly
chosen configurations X. For each configuration the potential energy
$U_p(X)$ is calculated and Z_{NVT} follows from:

$$Z_{NVT} \approx \frac{V^N}{K} \sum_{i=1}^K e^{-U_p(X)/kT} \qquad [5.23]$$

For dense systems such as liquids, this method requires an extremely
large number of configurations to be generated. This is because the
overwhelming majority of configurations will contain overlapping
molecules. These configurations contribute virtually nothing to the
partition sum (overlap of electron clouds means very high $U_p(X)$ and
thus very small Boltzmann factor in Eq. (5.23)).

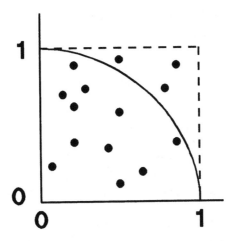

Figure 5.2. Monte Carlo approach to measuring the value of π/4 as the fraction of randomly chosen points that fall within the circle.

This complication can be circumvented by a 'less random' choice of configurations. In these so-called importance sampling techniques, only those "random" configurations are constructed that contribute significantly to the partition function. The principle of the method (also known as the Metropolis[9] algorithm) is as follows.

First a starting configuration is constructed. Then the following sequence of steps is executed:

1. A random departure of the configuration is chosen, for example by moving one or more particles.

2. The energy difference ΔU_p between the new and the old configuration is calculated: $\Delta U_p = U_p^{new} - U_p^{old}$.

3. If $\Delta U_p < 0$, the new configuration is 'accepted'. If $\Delta U_p > 0$, the new configuration is accepted with a probability $\exp(-\Delta U_p/kT)$.

4. The procedure starts again at step 1 with the new configuration if it was accepted or the old configuration if the new one was not accepted in the step 3.

These sequence of trials is repeated many times until equilibrium is reached. The criterion could be that the total energy of the system becomes constant, apart from fluctuations. Then step 1 to 4 are repeated again many times and each accepted configuration is counted as one contribution to the configurational partition sum.

The trick is that this method generates a chain of configurations that automatically have a Boltzmann distribution of energies. Stated differently: instead of choosing configurations randomly and weighting them with the Boltzmann factor, configurations are chosen with a probability given by the Boltzmann factor and weighted equally. The two most important (related) concerns of the method are: is the system ergodic such that all configurations are potentially sampled and what is the influence of the particular choice of the initial configuration? These concerns are usually tackled by repeating the Monte Carlo simulations with different initial configurations and showing that the results are the same.

Molecular dynamics simulations also generate a chain of configurations which have a Boltzmann distribution of energies. The main advantage of Monte Carlo methods is that the successive configurations may be much "further apart". Phase space is sampled with seven-leagers which diminishes the danger of sampling too small a phase space volume. The price that is paid is that time has no meaning anymore.

The above discussion was dealing with the canonical ensemble. Similar methods can be formulated for other ensembles. For systems involving phase equilibria, there is the complication of the interface between the two phases. This is circumvented by the so-called Gibbs Monte Carlo method.[10] The method uses two separate simulation boxes with a total fixed volumes $V = V_1 + V_2$ and fixed total number of particles $N = N_1 + N_2$. Standard Monte Carlo simulations are executed simultaneously in both boxes. In addition there are combined attempted volume changes in which one box changes with a volume ΔV while the other box changes in the opposite direction $(-\Delta V)$ so as to keep the total volume constant. There are also attempted moves of an arbitrary particle from one box, where it is distracted to a random point in the other box. Both volume changes and particle transitions are accepted or rejected according to the standard Metropolis algorithm.

The volume and number of particles in each box change during this process from the arbitrary initial configuration to one that is characteristic of the two coexisting phases. In equilibrium, the pressure and chemical potentials are equal in both simulation boxes.

5.5 SIMULATION OF SMALL SYSTEMS

Computer simulations are always limited to a finite number N of interacting species (atoms, groups of atoms, molecules). The storage capacity of the computer may impose limits on N. Also, a simulation requires the calculation of the $\sim N^2$ interaction energies and forces for each time step. The speed of the computer thus also imposes limits on the size N of the system. Many ingenious tricks have been developed to reduce the number of explicit calculations. Nevertheless, the speed of the computer is usually the limiting factor for the size N that can be managed.

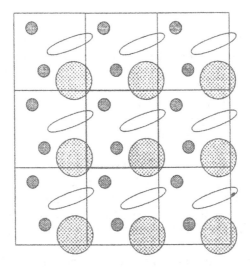

Figure 5.3. Periodic boundary conditions imposed on the central simulation box.

Two important complications of using finite size systems are: 1. surface effects of the boundaries and 2. interaction potentials that extend over the entire system. These problems are usually tackled by 1. applying periodic boundary conditions and 2. cutting off the potentials.

5.5.1 PERIODIC BOUNDARY CONDITIONS

Imposing periodic boundary conditions implies that the simulation box, say a square, containing the particles is replicated to form a small lattice of 9 identical boxes, see Figure 5.3. Imagine a molecule leaves the simulation box through, say, the right edge. Then its image in the box at the left edge of the simulation box, also leaves its box and enters the simulation box through the left edge. In this way a quasi continuum is simulated.

Usually, a spherical cut-off of the potential is applied where the maximum interaction distance is half the physical size L of the system.

Other tricks must be used when dealing with long range forces such as electrostatic interactions. Although some unwanted effects of periodic boundary conditions have been observed, particularly manifestation of anisotropy in isotropic systems,[11] the common experience is that periodic boundary conditions have little effect on the equilibrium thermodynamic properties.[1] For high molecular weight poly-

mers that may leave and re-enter the simulation box, complications
may occur.

5.6 SIMULATIONS OF POLYMERIC SYSTEMS

Computer simulations of polymeric systems with a focus on the typical
sorts of problems covered in this book are almost exclusively Monte
Carlo simulations. A broad distinction can be made between simulations of a single chain (in a solvent) and simulations of a more or less
dense system of many chains. Another broad distinction can be made
between simulations on a lattice and simulations in a continuum.

5.6.1 SINGLE CHAIN SIMULATIONS

The computationally most simple simulations are single chain simulations on a lattice. The total number of configurations on a lattice is
a finite number which greatly facilitates the statistics.

One of the first Monte Carlo studies of the configurations of a
'restricted random walk' (the term self-avoiding walk (SAW) still had
to be invented, the notion is the same: no crossing or doubling back)
on a lattice is due to Rosenbluth and Rosenbluth[12] in 1955. Their
procedure of generating and weighting SAWs is still used today. The
method will be illustrated on a 2 dimensional cubic lattice but can
straightforwardly be applied to any lattice. A self-avoiding walk of N
segments is build as follows: 1. the first link is placed from the origin
to $(x,y) = (1,0)$ and the chain is given a weight $W = 1$. For the second
link there are 3 possibilities and one is chosen at random. This process
continues until a situation occurs where one of the three options is
occupied by a previously filled-in position. Then one of the n ($0<n<2$)
remaining possibilities is picked at random and the weight of the
polymer chain is multiplied by $n/3$. This process is repeated until N
segments have been placed. The resulting chain with characteristics
such as the end-to-end distance, number of interchain contacts, persistence length, etc. is counted with its calculated weight in the final
statistical averaging of a large number of chains. A chain that ends-up
in a situation where all options are occupied is terminated and ignored
($W = 0$). Figure 5.4 shows an example of 13 segment SAW with its
corresponding weight and also of a chain that is trapped.

Rosenbluth and Rosenbluth calculated end-to-end distances R_{ee}
and the total number of possible SAW configurations, using this
method. They found the scaling relations $R_{ee} = A \, N^{\nu}$ and $\Omega_{SAW} = B \, z'^{N}$
with:

$A = 0.988$	$\nu = 0.725$	$B = 0.713$	$z' = 2.661$ in 2 dimensions and
$A = 1$	$\nu = 0.61$	$B = 0.292$	$z' = 4.705$ in 3 dimensions.

These values are very close to modern estimates. Half a year earlier,
Wall, Hiller, and Wheeler[13] published a similar study, though using

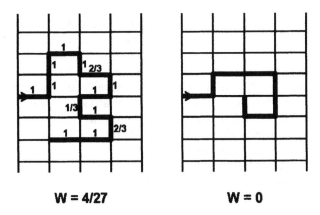

W = 4/27   W = 0

Figure 5.4. Example of the weighing of a 13 segment self-avoiding walk configuration (left) and a trapped configuration with zero weight (right).

a somewhat different technique for generating the chains. Another difference was that in the latter study bond angles on the cubic lattice were always 90°. They found an exponent $v = 0.61$ for both a cubic and a tetrahedral lattice, in perfect agreement with the results of Rosenbluth and Rosenbluth.

Since then, a large number of Monte Carlo type computer simulation studies of the behavior of a single chain have been published.

5.6.2 SIMULATIONS OF MANY CHAINS

Many authors have used Monte Carlo simulations to asses the validity of the Flory-Huggins theory and its corrections. The Flory-Huggins expression for the free energy of mixing has been derived on a lattice though the role of the lattice is very limited as discussed in Chapter 3. For this reason and ease of computation most Monte Carlo studies in this field have also been conducted on a lattice. Particular attention has been paid to the combinatorial entropy term of an athermal polymer solution. Such a solution is modeled as a collection of self and mutually avoiding chains. The volume fraction of the polymer determines the fraction of sites occupied by polymer segments. The remaining sites may be seen as solvent molecules but equally well as vacancies or free volume. In this sense there is a correspondence between the critical solution temperature of the polymer solution and the critical (liquid-vapor) point of the polymer.

One of the first studies of this kind was conducted by Bellemans and de Vos[14] in 1973. A 3 dimensional simple cubic lattice was filled with N self and mutually avoiding chains of r segments until a fraction Φ of the lattice sites was occupied by chain segments. Then the system was subjected to random Brownian motion for a certain period. The system configuration at the end of this process was sampled by random trials to insert another (randomly generated) chain into the system (removing 'solvent' segments). A successful trial is one in which none of the new segments overlap with existing chain segments. Clearly, the number of trials needed to obtain a successful trial increases horrendously for $\Phi \rightarrow 1$. These methods turn out to be only practical for $\Phi < 0.8$. One thus obtains a Monte Carlo estimate for the insertion probability P_{ins} of an extra chain which may be compared with theoretical expressions.

In the mean field model the insertion probability of r segments is simply the product of the insertion probabilities of 1 segment which in turn is simply given by the fraction $(1 - \Phi)$ of unoccupied sites:

$$P_{MF} = (1 - \Phi)^r \qquad [5.24]$$

In Chapter 3, the so-called Huggins correction was discussed (see Eq. (3.41)) which, to a certain extent, accounts for the correlation in site occupancies due to the connected character of the polymer chains. The "Huggins corrected" insertion probability P_{HC} is given by:

$$P_{HC} = (1 - \Phi)^r \left(\frac{r}{r - \dfrac{2}{z}(r - 1)\Phi} \right)^{r-1} \qquad [5.25]$$

which is always larger than P_{MF}.

Bellemans and de Vos found that for their systems of 6, 10, 20, and 30-mers, P_{HC} is a good approximation and significantly better than P_{MF}. As a by-product these authors also evaluated the mean square end-to-end distance $<R_{ee}^2>$ of the r-mers as a function of Φ. They found that the quantity $<R_{ee}^2(\Phi)>/<R_{ee}^2(0)>$ decreases almost linearly with Φ, steeper with larger n. They made the rough fit:

$$\frac{<R_{ee}^2(\Phi)>}{<R_{ee}^2(0)>} \approx 1 - 0.04n^{0.7}\Phi + \dots \qquad [5.26]$$

As discussed in Chapter 3, one would expect that for $\Phi \rightarrow 0$, the chains behave as self avoiding random walks, i.e, $R_{ee} \sim n^{0.6}$, while for $\Phi \rightarrow 1$, the chains should behave as random walks, i.e, $R_{ee} \sim n^{0.5}$. Consequently:

$$\frac{<R_{ee}^2(1)>}{<R_{ee}^2(0)>} \propto n^a \qquad [5.27]$$

with a = 0.2. Lack of data at Φ close to 1 prohibits testing of this prediction but the right trend is indeed observed. In a later paper, the same authors[15] reported on a more detailed Monte Carlo study of the scaling behavior of $<r^2>$ with the number of bonds n, using a more efficient algorithm. They found for the extrapolation $\Phi \rightarrow 1$:

$$<r^2> = (1.20 \pm 0.03)n^{1.075 \pm 0.010} \qquad [5.28]$$

Admitting that the chains considered may be too short ($n \leq 30$), this result was seen as an indication that the chains are not ideal and the intra- and intermolecular interactions do not exactly balance each other.

Later, Wall and Seitz[16] found that bulk ($\Phi \rightarrow 1$) polymer dimensions were very well described by what was called second order random walks. These are random walks without immediate self-reversal (loops of two bonds). They argue that indeed chain lengths n<30 are insufficient to observe $n \rightarrow \infty$ behavior.

A rotational isomeric state model of a polymer chain was used by Curro[17] in an early Monte Carlo study of the chain configuration as a function of chain density. The chain segments interacted via hard sphere potentials. At low concentrations, the chains are isolated. Since there are only hard sphere potentials this is equivalent to an athermal solvent. It was shown that when the chains start to overlap, the average chain dimension drops sharply to what Curro refers to as the random flight result.

Even in 1982, the issue of the Flory theorem about the ideality of chains in the bulk is still alive. Olaj and Lantschbauer[18] developed a novel technique that enabled Monte Carlo simulations of systems with $\Phi = 1$ (without extrapolation). This was achieved by allowing relaxation processes that involve breaking and combining chains. The price to pay is that a monodisperse (single chain length) system cannot be used. This need not be a serious problem as practical polymeric systems are always polydisperse and the polydispersity is relatively small. The simulations (n = 50) gave a quite convincing demonstration that indeed bulk polymer chains behave as random walks.

Okamoto[19] made an interesting comparison between insertion probabilities found in lattice systems as compared to continuum systems. It was found that P_{HC} performed better than P_{MF} in lattice systems. Interestingly, the reverse was true for continuum systems. Here the mean field expression was better. It must be noted that the study was restricted to 4, 5, and 6 mers in dilute and semi dilute concentrations.

Cifra et al.[20] studied the distribution of interactions in 2 dimensional binary polymer mixtures. The authors observed from their Monte Carlo simulations that even in athermal blends, the mean field theory overestimates the number of unlike contacts. This is attributed to the chain connectivities with effectively lead to screening of the interactions. This effect is particularly strong in 2 dimensional systems. As expected, the number of unlike contacts increases in case of favorable A-B interactions and decreases in case of unfavorable A-B interactions, i.e., respectively negative and positive exchange energy.

An extensive comparison of lattice theories with Monte Carlo simulations have been made by Madden et al.[21] The standard Flory theory was found to give good results at high volume fractions but a poor prediction of the critical temperature. Good agreement was found with extended mean field theories related to the n = 0 vector model[22] (see Chapter 4).

Sariban and Binder[23] made a study of the critical properties of lattice polymer fluids. They used a symmetric system (both polymers equally long) which allows transformation of a chain into a chain of the other type as potential Monte Carlo moves (in a grand canonical ensemble). An intriguing observation in this study is that the radii of the minority component chains were systematically smaller than the radii of the majority component along the coexistence curve.

Many interesting results have already been obtained from simulations of polymeric systems. A few of them have been discussed above. Nevertheless we believe it is save to say that the simulation techniques of realistic (chemically detailed and high molecular weight) polymeric systems have not yet reached a mature status. A major obstacle still is that realistic polymer liquids contain very large chains with a high density. Both aspects make Monte Carlo moves difficult. A large portion of the efforts in this area is therefore focused on developing better sampling techniques. Recent examples are the configurational bias technique[24] and extensions[25] and the calculation of 'segmental' chemical potentials.[26] As these methods develop and computers become faster, we will see a growing number of valuable contributions to polymer science from computer simulations.

REFERENCES

1.   M. P. Allen and D. J. Tildesley, *Computer Simulation of Liquids*, Oxford University Press, Oxford, 1989.
2.   R. J. Roe ed., *Computer Simulation of Polymers*, Prentice Hall, Englewood Cliffs, NJ, 1991.
3.   K. Binder, *Colloid Polym. Sci.*, **266**, 871 (1988).
4.   K. Kremer and G. S. Grest, *Computer simulations of polymers*, Prentice Hall, Englewood Cliffs, NJ, 1991.
5.   W. Moore, *Physical Chemistry*, Longman, London, 1972.
6.   D. McQuarrie, *Statistical Mechanics*, Harper & Row, NY, 1976.
7.   P. G. de Gennes, *Scaling Concepts in Polymer Physics*, Cornell University Press, Ithaca, NY, 1985.
8.   C. Kittel, *Introduction to Solid State Physics*, Wiley, New York, 1976.
9.   N. Metropolis, A. W. Rosenbluth, M. N. Rosenbluth, A. H. Teller, and E. Teller, *J. Chem. Phys.*, **21**, 1087 (1953).

10. A. Z. Panagiotopoulos, *Mol. Phys.*, **61**, 813 (1987).
11. R. W. Impey, W. G. Madden, and D. J. Tildesley, *Mol. Phys.*, **44**, 1319 (1981).
12. M. N. Rosenbluth and A. W. Rosenbluth, *J. Chem. Phys.*, **23**, 356 (1955).
13. F. T. Wall, L. A. Hiller, and D. J. Wheeler, *J. Chem. Phys.*, **22**, 1036 (1954).
14. A. Bellemans and E. de Vos, *J. Polymer Sci. Symp.*, **42**, 1195 (1973).
15. E. de Vos and A. Bellemans, *Macromolecules*, **8**, 651 (1975).
16. F. T. Wall and W. A. Seitz, *J. Chem. Phys.*, **67**, 3722 (1977).
17. J. G. Curro, *J. Chem. Phys.*, **61**, 1203 (1974).
18. O. F. Olaj and W. Lantschbauer, *Makromol. Chem. Rapid Commun.*, **3**, 847 (1982).
19. H. Okamoto, *J. Chem. Phys.*, **64**, 2686 (1976).
20. P. Cifra, F. E. Karasz, and W. J. MacKnight, *Macromolecules*, **21**, 446 (1988).
21. W. G. Madden, A. I. Pesc, and K. F. Freed, *Macromolecules*, **23**, 1181 (1990).
22. J. Dudowitz, K. F. Freed, and W. G. Madden, *Macromolecules*,
23. A. Sariban and K. Binder, *J. Chem. Phys.*, **86**, 5859 (1987).
24. B. I. Siepmann and D. Frenkel, *Mol. Phys.*, **75**, 59 (1992).
25. J. J. de Pablo, M. Laso, and U. W. Suter, *J. Chem. Phys.*, **96**, 6157 (1992).
26. S. K. Kumar, I. Szleifer, and A. Z. Panagiotopoulos, *Phys. Rev. Lett.*, **66**, 2935 (1991).

Chapter 6

EXPERIMENTAL FINDINGS

6.1 INTRODUCTION

In this section we will discuss experimental techniques and results as well as the validation of thermodynamic model descriptions related to the molecular miscibility of polymer solutions and high molecular weight polymer mixtures. The so-called compatible mixtures, which are actually immiscible or (semi-) crystalline but macroscopically homogenized using appropriate processing or compatibilization techniques, will not be discussed, because we want to keep the arguments limited to the thermodynamic miscibility on a molecular scale. Hence, we will present and discuss mainly those experimental techniques that are quantitatively related to the thermodynamic phase behavior of polymeric mixtures, in particular scattering techniques. For a more complete overview on compatible mixtures and related experimental techniques the interested reader may want to study Olabisi et al.[1]

In everyday practice of polymer blending, immiscibility or partial miscibility is perhaps of more relevance than miscibility. This is because the relevant mechanical properties of blends are enhanced mainly through phase separated domains which, with the aid of appropriate processing technologies, may transform into molecularly aligned and reinforced structures. On the other hand, molecular miscibility is relevant with regard to the control of translucency of films or the control of glass transition temperatures when impact is an issue. A basic understanding of thermodynamics as well as dynamics of phase-separation is a prerequisite to control of relevant parameters that determine the ultimate domain size and morphology of partially miscible or even immiscible blends. Such understanding is undoubtfully useful as well for the control of polymerization processes in solution. Hence, we trust that the more fundamental thermodynamic approach fits a purpose.

6.2 EXPERIMENTAL TECHNIQUES

6.2.1 SCATTERING TECHNIQUES

The most direct way to measure the thermodynamic miscibility of polymer solutions or mixtures is via scattering techniques, namely light, neutron, or X-ray scattering. The underlying reason is that such scattering is caused by microscopic concentration fluctuations in a mixture. Let us assume that we have a microscope with a sufficiently large magnification to actually "see" polymer chains moving in a mixture of, for example, red and yellow chains. At low magnification, the mixture will look orange: there is a homogeneous distribution of the chains, and an average, "orange" concentration $<c>$ is seen. At sufficiently large magnifications however, we will observe red and yellow spots that are continuously moving around, thereby changing in size and shape. We will observe that spots of one color grow, shrink, and even disappear in favor of the other color: there appears to be a well defined equilibrium. This equilibrium is a thermodynamic equilibrium. In a thermodynamically miscible mixture, concentration fluctuations grow because of random thermal motion of the polymers. But the free energy of mixing, favoring molecular miscibility of the red and yellow species, acts as a restoring force, preventing the fluctuations from growing too large. This equilibrium is described in a quantitative way in the fluctuation theory of Einstein and Smoluchowski.[2-4] This theory links the probability distribution of concentration fluctuations δc to the free energy curvature:

$$<\delta c^2> = k_B T \frac{1}{\partial^2 \Delta G_M / \partial c^2} \qquad [6.1]$$

This equation states that the smaller the free energy curvature (with respect to concentration), the smaller the restoring force, the larger the (mean square average of) concentration fluctuations, *and vice versa*.

Note that Eq. (6.1) has a solution only if the free energy curvature with respect to concentration is positive. In other words, the mixture should be either in the stable or the metastable state. As discussed in detail in Chapter 2, if the thermodynamic state of a mixture is stable, not only the free energy curvature around the average concentration $<c>$ is positive, but the free energy itself is at a minimum at $<c>$. In the stable state, the mixture is thermodynamically miscible and in the homogeneous one-phase region. In this region, the free energy curvature can be measured directly via scattering techniques, allowing comparison with theoretical model predictions. It will be shown below that in the special case of a stable dilute polymer solution, scattering allows determination of radii of gyration, molecular weights, and osmotic second virial coefficients, via which the Flory-Huggins interaction parameter χ can be determined.

Eq. (6.1) is also valid if the thermodynamic state of the mixture is metastable, that is, if the concentration $<c>$ and/or pressure p or temperature T are such that the mixture is in between the binodal and the spinodal curve. In the case of mixtures, this state is perhaps best characterized as "supersaturated". For example, a tube filled with a metastable dilute polymer solution would look quite homogeneous, transparent and therefore thermodynamically miscible. A bit of shaking would render the mixture completely "cloudy" and after a few seconds, a cloudy polymer-rich phase would appear at the bottom and a transparent solvent-rich phase would appear at the top of the tube. This example illustrates the essence of metastability: although the free energy curvature is positive around $<c>$, the free energy itself is not at a minimum at $<c>$. Coexisting phases with compositions c^I and c^{II} exist such that the total free energy is lower than the free energy at $<c>$, see also Figure 2.1.

Although, the principle of metastability states that a more stable thermodynamic state exists, it does not specify how this situation will be reached. This is a matter of mechanics rather than thermodynamics. The actual mechanism is that of nucleation and growth, which is the process of generating within the metastable phase the initial nuclei of the more stable coexisting phase. In principle, these nuclei can be formed in two ways. In the homogeneous way, there is a small, but finite, possibility that a random concentration fluctuation around $<c>$ has sufficient amplitude to activate the mixture over the free energy barrier separating the coexisting phases with concentrations c^I and c^{II}, see Figure 2.1. Note that the probability of nucleation increases strongly with the degree of supersaturation, because the free energy curvature evaluates to zero when the spinodal curve is approached, thereby causing an enormous increase (even a divergence at the spinodal) of concentration fluctuations.

In the heterogeneous way, the necessary activation energy is brought about by external factors such as inhomogeneities, contaminations, surface structure of walls, external energy input (shaking), etc. Once one or more nuclei are formed, irrespective of whether these are homo- or heterogeneous, free energy is lowered locally and phase separation phenomena will proceed until macroscopic phase separation is reached and the mixture is in thermodynamic equilibrium. Phase separation in the metastable region via the nucleation and growth mechanism forms the basis of cloud point measurements, to be discussed below.

Some special consideration should be given to the loci of points where the second derivative of the free energy evaluates to zero, the spinodal. As discussed above, when the spinodal is approached from the metastable phase, concentration fluctuations increase until phase separation sets in via the nucleation and growth mechanism. These enhanced concentration fluctuations lead to an enormous increase of the scattering intensity close to the spinodal, the so-called critical opalescence,[5] a phenomenon that is common to all types of mixtures, not necessarily polymeric mixtures. Critical opalescence makes spinodals experimentally accessible. This increase (and ultimately, divergence) of scattering intensity can already be a sizable effect even in the miscible stable state if binodal and spinodal are not too far removed. The effect is strongest under critical conditions, that is, when the spinodal is approached in that area of the phase diagram where binodal and spinodal come together, the critical point.

Note that at the point where binodal and spinodal come together (see also Figure 2.1) there is no metastable region. Phase stability directly swaps from stable (positive free energy curvature) to unstable (negative free energy curvature). Hence, the critical point is the only point in the phase diagram via which the unstable region is experimentally accessible without interference of the nucleation and growth mechanism of the metastable region. In the unstable region there is no restoring force for concentration fluctuations: they continue to grow throughout the mixture without any free energy penalty interfering. Most interesting is the early stage of this growth mechanism, called spinodal decomposition. It will be shown below that mechanical laws prescribe that in the unstable phase concentration fluctuations initially develop as sinusoidal modulations with a well-defined wavelength. In other words, in a binary mixture a very regular and interwoven two-phase structure develops. In polymeric mixtures, the time scale of such development is relatively slow and experimentally accessible. The regularity of the structure depends on the exact location in the unstable phase, i.e., the distance from the spinodal. Hence, spinodal decomposition is in principle another method to determine the location of the critical point and, under well defined quenching conditions, the spinodal curve, as will be discussed in more detail below. In the later stage of spinodal decomposition, the modulated structure coarsens to a more dispersed structure. Obviously, the final stage is the stable equilibrium stage, where full macroscopic phase separation has developed.

6.2.1.1 Cloud point methods

The nucleation and growth mechanism is the basis of a simple scattering technique employed in the study of polymeric phase behavior, namely the cloud point technique. A stable homogeneous mixture, either a polymer solution or a polymer mixture, is transparent. Thermodynamic mixing implies that polymers and/or solvent are mixed on a molecular scale, so that the refractive index for visible light is homogeneous on a microscopic level. Given a homogeneous mixture, the metastable region can be entered by changing temperature, pressure, or composition of the mixture. Then there is a finite probability that the nucleation and growth mechanism sets and a two phase structure develops. If the mixture is asymmetric (low concentration of component A and high concentration of component B), droplets of A-rich phase will grow in a matrix of B-rich phase. Once these droplets have grown to the same order of magnitude as the wavelength of visible light (400 ÷ 800 nm), the refractive index difference between the two phases becomes important and the mixture becomes turbid, less transparent or "cloudy". In the case of a more symmetric mixture, a more interwoven structure develops, but a cloudy state develops in a similar manner. The cloud point corresponds to the transition from the transparent to the cloudy state. It depends on experimental conditions. In general, a higher degree of metastability (supersaturation) can be obtained at relatively high cooling/heating rate. Therefore, if the cloud point is to represent the real binodal in the case of binary mixtures, experiments at various temperature-pressure-composition variation rates need to be done and extrapolation to the state of zero rate is to be preferred. Disappearance of the cloud point upon reversal of the temperature-pressure-composition variation should be observed as well. If appearance and disappearance of the cloud point coincide upon a full cycle, one is certain that a thermodynamically relevant binodal point is determined.

Due to the relatively low viscosity of dilute polymer solutions, cloud point measurements on such systems are fairly easy and therefore often reported. Binodals of what is probably the most studied model polymer solution, polystyrene (PS) in cyclohexane (CH) are shown in Figure 6.1, after Saeki et al.[6] To obtain such results, several solutions were prepared in the concentration range from 1% to 25% PS and flame sealed under dry nitrogen gas in cylindrical glass cuvettes. A glass sphere was inserted in each cuvette to stir the solution with the aim of inducing heterogeneous nucleation. With the aid of a simple He-Ne laser beam (see below) and by extremely slow variation of temperature cloud points could be determined with an accuracy of 0.05 degrees. Note that both phase separation upon cooling (Upper Critical Solution Temperature, UCST) and phase separation upon heating (Lower Critical Solution Temperature, LCST) are observed. Note also that the phase diagrams are skewed towards the polymer poor phase and that this skewness increases with molecular weight, as predicted by Flory-Huggins theory, see also Eq. 3.58.

The critical reader may wonder which factors determine the actual concentration range investigated. Although, the ranges may differ from mixture to mixture, there are some general comments to be made. There is virtually no limitation on the low concentration side. If one has a sufficiently strong laser beam, one is able to detect even

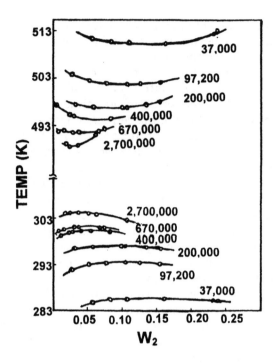

Figure 6.1. Phase diagrams (cloud point curves) of polystyrene in cyclohexane for polystyrene of indicated molecular weight. Adapted, by permission from S. Saeki, N. Kuwahara, S. Konno, and M. Kaneko, *Macromolecules*, **6**, 246 (1973)).

the slightest change in refractive index even in an extremely dilute polymer solution. There is however a limitation on the high concentration side. One has to realize that already at relatively low polymer concentrations, polymers start to "overlap" each other. Physics behind such overlap is described by de Gennes.[7] For the present purpose it is sufficient to know that the concentration at which chains overlap is a function of molecular weight and interaction parameter, and is in magnitude similar (but not necessarily equal) to the critical concentration of the solution. Hence, in the high concentration range (larger than typically 10% for relatively low molecular weights and 1% for high molecular weights), mixtures are far above their respective overlap concentrations. They are in the "gel" state. Due to the connectivity of polymer chains, the viscosity becomes very large and the mixture becomes less transparent due to the refractive index contrast between the large polymeric network and the solvent. Due to the high viscosity of the mixture the nucleation and growth mechanism becomes extremely slow and difficult to distinguish from already turbid stable phase. The lower the viscosity (and hence, molecular weight),

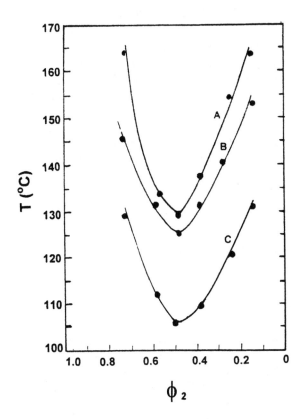

Figure 6.2. Cloud point curves of polypropylacrylate (A), polybutylacrylate (B), and polypentylacrylate (C) with PVC. (Adapted, by permission, from C.K. Sham and D. Walsh, *Polymer*, **28**, 957 (1987)).

the larger the concentration range that can be investigated, see Figure 6.1.

In the rare case that high molecular weight polymeric mixtures have a relevant region of thermodynamic miscibility, their high viscosity makes it difficult to determine the location of cloud point curves. Actually, the relatively slow kinetics in these mixtures makes it difficult to assess whether the mixture is in a thermodynamic equilibrium state at all. Hence, the preparation route of high molecular weight model polymer mixtures is essential. A reliable route, often used, is to cast a thin film of mixture from a common good solvent. The film is put in a hot-stage oven allowing observation of phase separation phenomena with the aid of a microscope or a laser beam. Various temperature cycles are carried out and cloud points are recorded. A result of such an experiment is given in Figure 6.2.

6.2.1.2 Spinodal decomposition

In the case of phase separation in the unstable phase there is no restoring force for concentration fluctuations. This is associated with the peculiar phase separating mechanism called spinodal decomposition (SD). It is a consequence of the mechanical laws behind microscopic concentration fluctuations of a certain amplitude and length scale around an average value $<c>$. The relevant microscopic diffusion equations were solved by Cahn.[9] The linearized solution describes the time and wavevector dependence of concentration fluctuations $\delta c(q,t)$ in terms of a general local free energy expansion around $<c>$:

$$\partial c(q,t) = \partial c(q,t=0) \exp \left(R_c(q)t \right) \qquad [6.2]$$

where the relaxation rate of concentration fluctuations R_c is given by:

$$R_c(q) = -M_{AB}q^2 \left(\frac{\partial^2 G^V}{\partial c^2} + 2\kappa q^2 \right) \qquad [6.3]$$

M_{AB} is the mobility between polymers A and B, $q = 2\pi/\lambda$ is the wavevector of the spatial composition fluctuations with wavelength λ. The term $2\kappa q^2$ (with κ an unknown positive constant) can be regarded as the free energy penalty that has to be paid for microscopic concentration fluctuations around the average value $<c>$ anywhere, but locally in the mixture. G^V is the free energy per unit volume. Note that in the thermodynamic limit $q \rightarrow 0$ local contributions to the free energy should disappear. This is indeed the case, because in this limit the term $2\kappa q^2$ can be neglected compared to $\partial^2 G^V/\partial c^2$ and the macroscopic diffusion equation

$$R_c(q) = -D_{AB}q^2 \qquad [6.4]$$

is recovered. The phenomenological diffusion coefficient D_{AB} is defined by:

$$D_{AB} = M_{AB} \frac{\partial^2 G^V}{\partial c^2} \qquad [6.5]$$

In the stable or metastable region of phase diagram, $\partial^2 G^V/\partial c^2$ is positive. Consequently, $R_c(q)$ is always negative and infinitesimal fluctuations cannot grow, but rather decay on a time scale given by $R(q)^{-1}$. In the unstable region, $\partial^2 G^V/\partial c^2$ is negative and hence there will be always a positive $R(q)$ for some value of q smaller than a critical value q_c. Therefore, any infinitesimal fluctuation with wavevector $q<q_c$ can grow at a rate $R_c(q)$, which will lead to a decay into the phase separated state via a special topological configuration. By setting Eq.

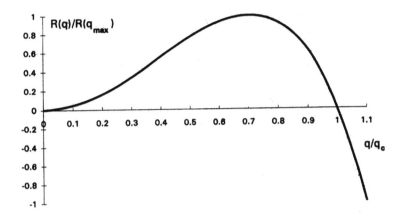

Figure 6.3. Typical shape of R(q) versus q.

(6.3) equal to zero it followed that any fluctuation with wavevector $q^2 < q_c^2 = (1/2\kappa)\partial^2 G^V/\partial c^2$ can grow. By differentiating Eq. (6.3) with respect to q it follows that $R_c(q)$ has a maximum at

$$q^2 = q_{max}^2 = \frac{1}{2}q_c^2 = -\frac{1}{4\kappa}\frac{\partial^2 G^V}{\partial c^2} \qquad [6.6]$$

q_{max} is the wavevector for which $R_c(q)$ grows most rapidly. The value of R_c at q_{max} is:

$$R(q_{max}) = \frac{M_{AB}}{8\kappa}\left(\frac{\partial^2 G^V}{\partial c^2}\right)^2 \qquad [6.7]$$

Note that q_{max} is controlled by thermodynamics only whereas $R(q_{max})$ also depends on mobility M_{AB}. A typical shape of the function R(q) versus q is shown in Figure 6.3

The crucial point in SD is that the diffusion coefficient D_{AB} is negative (and thus, R(q) positive) in the unstable region of the phase diagram, which causes the amplification of long wavelength fluctuations (Cahn characterized SD as "uphill diffusion"). At shorter wavelengths (larger q values) the amplification is increasingly more compensated for by the gradient term $-2\kappa M_{AB}q^4$, which gives rise to the distinct maximum $R(q_{max})$. Thus, the existence of a specific wavelength q_{max}, at which concentration fluctuations increase most rapidly, can be interpreted as a wavevector dependent competition between a thermodynamic driving force, by which particles A and B move against their respective concentration gradients, and a restoring force resulting from these gradients as in "ordinary" diffusion.

Preferential amplification of concentration fluctuations at q_{max} leads to a topological structure which is typical for SD, namely a three-dimensional cocontinuous interpenetrating network of an A-rich and an A-poor phase separated by typical distances of λ_{max}, provided that volume fractions of A and or B phase are not lower than some 15%.

The linearized theory of SD is only valid at short times. At larger time scales, more higher order effects become important and the description given above fails. The so called "late stage growth" of SD is not yet fully understood. Nevertheless, also in case of SD the final stage is macroscopic phase separation into the thermodynamic equilibrium state. Hence, the growing cocontinuous structure will gradually change into a more coarse coalescing droplet type structure and in the end two macroscopic phases will be left.

Dynamics of SD can be studied by rapidly quenching a sample, e.g., a polymer blend prepared as a thin film, from a temperature in the stable state to a temperature in the unstable state and measuring the time development of the scattered intensity directly from the start of the quench. It is expected that after a certain period of time a distinct scattering maximum will appear and grow at a rate proportional to $R(q)$. This process can be measured by continuously probing the angular dependence (and therefore the q dependence, see also below) of the scattered intensity as the SD mechanism proceeds. For example, from a study of Hashimoto et al.,[10] it was found that in a PS/PVME blend a maximum may grow without changes its position for about 1 hour, see Figure 6.4.

The interesting consequence of the SD technique is that thermodynamic quantities defined in Eq. 6.7 can be determined directly using a set of relatively simple quenching experiments.

6.2.1.3 Light scattering

In the quantitative study of the thermodynamics of polymeric mixtures scattering techniques are essential. As indicated above, the scattering mechanism directly probes concentration fluctuations and therefore free energy curvature. We will present the general mechanisms behind light scattering here. Most of these will also be applicable to neutron- and X-ray scattering. In a typical light scattering experiment, a monochromatic laser beam passes through a medium

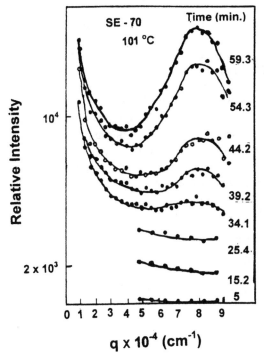

Figure 6.4. Change of the scattered intensity profiles with time during a course of SD in a PS/PVME blend. (Adapted, by permission, from T. Hashimoto, J. Kumaki, and H. Kawai, *Macromolecules*, **16**, 641 (1983)).

with dielectric constant ε. This medium can be any substance that can be polarized. The physical idea is that the laser light induces an oscillating dipole field in the medium so that radiation will be emitted, which causes a (usually small) attenuation of the primary beam. By looking at the average scattering intensity as a function of the direction in which the light is scattered, information on spatial interference effects between the local dipoles induced in the medium is obtained. In other words, one can look at spatial correlations which have dimensions in the order of the wavelength of the laser light, which is typically in the order of 500 nm. In the special case of a dilute polymer solution for instance, the scattering pattern is related to the size of single polymer coils.

The scattering intensity will fluctuate very rapidly due to the thermal motion of the molecules. If detection is fast enough the time dependence of such fluctuation can be studied. In particular, by measuring the so-called autocorrelation function of the scattered intensity, one can determine in how far the light scattering pattern is correlated in time, which means that one can measure the probability of a particle (or polymer segment) to be at a certain place at time zero, it will still be at the same place a certain time later. For mixtures this is actually a method to measure diffusion coefficients.[11]

To become more quantitative, we have to study the electromagnetic Maxwell equations that govern the physics behind the scattering process. We consider the usual geometry for the light scattered from a cylindrical cuvette, where the incident beam is vertically polarized and the light is detected in the horizontal plane perpendicular to the polarization vector of the incident beam. In this geometry, light is scattered from a macroscopic volume V which is the intersection of the primary laser beam and the opening angle of a detector. Light is scattered from this volume in all directions (in a full solid angle 4π). The general formula for the angle and time dependent scattering intensity $I(q(\Theta),t)$ is given by:[12,13]

$$I(q(\Theta)) = \frac{kV^2}{R} <\delta\varepsilon(q)^2>I_o \qquad [6.8]$$

where K is defined by:

$$K = \frac{1}{16\pi^2}\left(\frac{2\pi}{\lambda}\right)^4 = \frac{\pi_2}{\lambda^4} \qquad [6.9]$$

I_o is the intensity of the primary laser beam and R is the distance from the macroscopic scattering volume to the detector. Formally, V is the volume element associated with the fluctuation in the dielectric constant $\delta\varepsilon$. The λ^{-4} term in Eq. (6.9), λ denoting the wavelength of the light, e.g., 514 nm, is peculiar for light scattering and comes from the nature of the Maxwell equations. It specifies that blue light is scattered much more intensely than red light: it causes the blue colors of the skies and oceans. In polymer mixtures, it causes the initial "bluish" haziness that can be observed in case of phase separation in the very early stage when the nuclei are still very small. In the later stage of nucleation and growth, when the amount and size of phase separating domains has grown, the so-called multiple scattering effects become important and generate a more wavelength independent scattering pattern: the mixture becomes white and turbid.

In Eq. (6.8), $q(\Theta)$ denotes the wavevector change in the scattering event:

$$q = 2q_i\sin\left(\frac{\theta}{2}\right) = 2\frac{2\pi}{\lambda_i}\sin\left(\frac{\theta}{2}\right) \qquad [6.10]$$

where $q_i = 2\pi/\lambda_i$ is the wavevector of the incident beam. This is the Bragg condition for scattering, which is valid for all scattering mechanisms (light, neutron, X-ray). It exactly specifies the wavevector q that gives rise to scattering at an angle Θ.

A practical quantity used in light scattering is the Rayleigh ratio $R(q)$, which is defined as the ratio of the scattered intensity to the primary beam intensity multiplied by R^2/V:

$$R(q) = \frac{I(q)R^2}{I_0 V} = KV<\delta\varepsilon(q)^2> \tag{6.11}$$

$R(q)$ denotes the scattering intensity per unit volume, scattered in a unit solid angle in the direction Θ and its dimension is meter^{-1}. It is directly related to the turbidity τ, which is the total scattering intensity taken over the complete solid angle Ω:

$$\tau = \int d\Omega(\theta,\phi)R(q(\theta))\sin^2(\phi) \tag{6.12}$$

where the $\sin^2(\phi)$ term comes form the symmetry of the dipole radiation tensor.[12,13] If the molecular weight of the solution is low (simple solutions, no polymers) and in the stable phase of the phase diagram (no long range correlations), the spatial correlation length of fluctuations is negligibly small compared to the wavelength of the laser light and the Rayleigh ratio has no angular dependence. It is isotropic with a value R_0. In this case it can be shown that Eq. (6.12) simplifies to:

$$\tau = \frac{(8\pi)}{3} R_0 \tag{6.13}$$

If the turbidity τ is sufficiently large, a sizable attenuation of both incident and scattered light may result. Let's take the typical experimental set-up in which scattering takes place at the center of a cylindrical cuvette with diameter d. Before entering the scattering volume, the scattering intensity will be attenuated by a factor $\exp(-\tau d/2)$. On leaving the cuvette, the now scattered light will be attenuated by the same factor. Hence, a measured Rayleigh ratio R_m is related to the real Rayleigh ratio R by:

$$R_m = R \exp(-\tau d) \tag{6.14}$$

For example, the "scattering length" τ^{-1} amounts to 3 km for pure benzene and 10 m for a 1% polymer solution in benzene.[13] With a typical diameter of about 10 mm, such attenuation is negligibly small, so that   "true" Rayleigh ratios can be measured. However, when the metastable region or even unstable region of phase diagram sd is approached, turbidity effects become important, as will be shown in more detail in Section 6.3.

In a mixture, fluctuations in the dielectric constant ε are coupled to fluctuations in concentration c and density ρ:

$$<\delta\varepsilon(q)^2> = \left(\frac{\partial\varepsilon}{\partial c}\right)^2 <\delta c^2> + \left(\frac{\partial\varepsilon}{\partial\rho}\right)^2 <\delta\rho^2> + \left(\frac{\partial\varepsilon}{\partial c}\right)\left(\frac{\partial\varepsilon}{\partial\rho}\right)<\delta\rho\delta c> \qquad [6.15]$$

In polymeric mixtures and solutions the contribution from density fluctuations is usually much smaller than that from concentration fluctuations. In addition, the contrast factor $\partial\varepsilon/\partial c$ is orders of magnitude larger than $\partial\varepsilon/\partial\rho$ because of the large difference in dielectric constant between the consituents. Therefore, in practice the contribution from density fluctuations can be neglected or corrected for as a small background term. The expression for the Rayleigh ratio now becomes:

$$R(q) = KV\left(\frac{\partial\varepsilon}{\partial c}\right)^2 <\delta c(q)^2> \qquad [6.16]$$

The dielectric constant ε is related to the refractive index n:

$$\varepsilon \equiv n^2, \quad \delta\varepsilon = \delta n^2 = 2n\delta n, \quad \left(\frac{\partial\varepsilon}{\partial c}\right) = 2n\left(\frac{\partial n}{\partial c}\right) \qquad [6.17]$$

The expression for the Rayleigh ratio thus becomes:

$$R(q) = KV4n^2\left(\frac{\partial n}{\partial c}\right)<\delta c(q)^2> \qquad [6.18]$$

This equation is connected to Eq. (6.1), which relates the mean-square value of concentration fluctuations $<\delta c^2>$ with the free energy curvature. The following thermodynamic limit is obtained:

$$\lim_{q->0} <\delta c(q)^2> \equiv <\delta c^2> = k_BT\left(\frac{\partial^2\Delta G}{\partial c^2}\right)^{-1} \qquad [6.19]$$

which gives:

$$\lim_{q\to0} R(q) \equiv R(0) = KV4n^2\left(\frac{\partial n}{\partial c}\right)^2 k_BT\left(\frac{\partial^2\Delta G}{\partial c^2}\right)^{-1} \qquad [6.20]$$

Since the free energy curvature is associated with the volume element V, Eq. (6.20) can be rewritten as:

$$R(0) = K4n^2\left(\frac{\partial n}{\partial c}\right)^2 k_BT\left(\frac{\partial^2\Delta G^V}{\partial c^2}\right)^{-1} \qquad [6.21]$$

where ΔG^V is the free energy per unit volume. Hence, by extrapolating the scattered intensity to zero scattering angle (zero wavevector), the free energy curvature is determined quantitatively, allowing for comparison with theoretical model predictions.

Eq. (6.21) is generally valid, irrespective of the concentration or phase stability of the mixture. However, a useful extrapolation to zero scattering angle is needed. To describe the angular dependence the static structure factor $S(q)$ is defined:

$$S(q) = \frac{R(q(\theta))}{R(q(0))} \tag{6.22}$$

Obviously, $S(q = 0) \equiv 1$.

For dilute polymer solutions all angular dependence of $S(q)$ comes from the interference within single polymer chains, and in that case $S(q)$ reduces to the single particle structure factor (sometimes also referred to as form factor) $P(q)$:[14]

$$S(q) = P(q) = 1 - \frac{1}{3}<R_g^2>_z q^2 + \dots \tag{6.23}$$

where R_g is the radius of gyration of a single polymer chain. This description is not valid for more concentrated solutions where polymer coils overlap and interact.

It can be seen form Eq. (6.21) that the zero-angle Rayleigh ratio diverges when the curvature of the free energy surface becomes zero (which defines the spinodal). In the case the spinodal is approached from the metastable phase of the phase diagram, this phenomenon is well-known as critical opalescence.[5] It is accompanied by long range concentration fluctuations with wavelength ξ, because the thermodynamic restoring force for these fluctuations vanishes close to the spinodal. Hence, close to spinodals the structure factor must be described in terms of long range correlations rather than as a single particle form factor. An analytical expression for the structure factor $S(q)$ close to spinodals has been derived by Ornstein and Zernike:[5]

$$S(q) = \frac{1}{1 + \xi^2 q^2} \tag{6.24}$$

Note that in expanded form Eqs. (6.24) and (6.23) have the same q-dependence provided that q is not too large. This implies that in the relevant q-range any $S(q)$, either single particle or more long range in nature, can be written in the Lorentzian profile of Eq. (6.24).

6.2.1.4 Dilute polymer solutions

In the special case of dilute polymer solutions, Eq. (6.21) can be written in a simplified analytical form allowing direct comparison with Flory-Huggins type thermodynamic models. At low concentrations, the free energy curvature can be expanded:[16]

$$\frac{1}{k_B T} \frac{\partial^2 \Delta G^V}{\partial c^2} = \frac{N_A}{c} \left(\frac{1}{M_w} + 2A_2 c + \ldots \right)$$ [6.25]

where A_2 is the osmotic second virial coefficient. Substitution of Eq. (6.25) into (6.21) gives the following relation:

$$\frac{K^* C}{R(0)} = \frac{1}{M_w} + 2A_2 c$$ [6.26]

where K^* is defined as:

$$K^* = \frac{4\pi^2}{N_A \lambda^4} n^2 \left(\frac{\partial n}{\partial c} \right)^2$$ [6.27]

Hence, from a relevant extrapolation to zero scattering angle weight-average molecular weight M_w and virial coefficient A_2 in dilute polymer solutions can be determined. Refractive index n and refractive index increment $\partial n / \partial c$ can be determined independently using differential refractometry.[13] The extrapolation to zero scattering angle is obtained by introducing the single particle form factor P(q) (Eq. (6.23)) into Eq. (6.21):

$$R(q) = R(0)P(q) = \frac{K^* c}{\frac{1}{M_w} + 2A_2 c} \left(1 - \frac{1}{3} <R_g^2>_z q^2 \right)$$ [6.28]

The relevant q range is chosen in such a way that $R_g^2 q^2 \ll 1$ and the higher moments in the form factor can be neglected. In this particular case, Eq. (6.26) obtains the following q-dependence:

$$\frac{K^* c}{R(0)} = \frac{1}{M_w} + 2A_2 c + \frac{<R_g^2>_z}{3M_w} q^2$$ [6.29]

which can also be written as:

$$\frac{K^* c}{R(0)} = \frac{1}{M_w} + \frac{<R_g^2>_z}{3M_w} \left(q^2 + \frac{6A_2 M_w}{<R_g^2>_z} c \right)$$ [6.30]

It is a useful procedure to make a plot of $K^* c / R(q)$ versus q^2+ (constant) × c, the so called Zimm-plot, where (constant) can be chosen

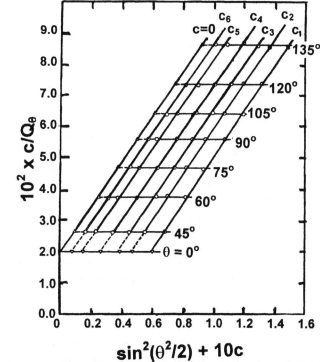

$$\text{sin}^2(\theta^2/2) + 10c$$

Figure 6.5. Zimm plot of polytetrahydrofuran in iso-propanol. (Adapted, by permission, from M.B. Huglin, **Light scattering from polymer solutions**, *Academic*, NY, 1972).

equal to the coefficient in Eq. (6.30), but can also be chosen arbitrary. As a matter of fact, (constant) will in practice be chosen such that the plot gives a grid of straight lines, both at constant c and q^2. It is the experimental task to measure the scattering intensity at various low concentrations and scattering angles, such that extrapolations to $q = 0$ and $c = 0$ are possible. The following parameters can be determined from a standard Zimm-analysis:

$$M_w = \text{intercept } (q = 0, c = 0), \quad <R_g^2>_z = 3M_w \text{ slope } (c = 0) \quad [6.31]$$

$$A_2 = \frac{1}{2} \text{ slope } (q = 0)$$

A typical example of a Zimm-plot is shown in Figure 6.5. In this particular case, data were obtained under Θ conditions, so that $A_2 = 0$.

6.2.1.5 Neutron scattering

Although the light scattering technique appeared useful in the investigation of polymer solutions, draw-back is that it requires a dust-free sample preparation procedure to allow for an accurate and quantitative extraction procedure from scattered intensity to free energy curvature. Such procedure is difficult to employ in the preparation of relatively high viscosity polymer blend samples. The neutron labeling technique (see below) proved a useful alternative here and hence SANS, although more intricate (and expensive !) than light scattering, is commonly used to study the thermodynamic behavior of high molecular weight polymeric mixtures.

There is no fundamental difference between light- and neutron scattering. When properly used, both methods give quantitative information on free energy curvature through the zero-angle scattering intensity. A relation similar to Eq. (6.21) is also obtained with neutron scattering. Main difference between both methods is basically in the proportionality constants in Eq. (6.21). In light scattering the electromagnetic field is coupled with concentration fluctuations via the dielectric constant of the medium or, in other words, the polarizability of polymer segments. In the case of neutron scattering, the relevant interaction is of the neutron-nucleus type. This leads to some technical differences.[16]

The source for light scattering is normally a (relatively simple) laser, with a typical (monochromatic) wavelength in the range of 500-600 nm. Neutrons however come from a more complicated source such as a nuclear reactor or a particle accelerator. Such neutrons are thermal in nature and have a wavelength in the range 0.5-5 nm, orders of magnitude shorter than laser light. In the investigation of high molecular weight polymeric mixtures the full angular scattering range between $20°$ and, e.g., $160°$ from the forward direction is used in light scattering in order to obtain a useful wavevector range (see Eq. (6.10)) and hence an extrapolation to zero scattering angle. Since the neutron wavevector is orders of magnitude larger than that of light, the angular scattering range investigated in neutron scattering needs to be orders of magnitude smaller to cover the same scattering vector range, relevant for polymeric systems, as in light scattering. Hence, scattered neutrons are normally detected in a small scattering range between a few tenths of degrees to a few degrees from the forward direction only. Such technique is referred to as **Small Angle Neutron Scattering (SANS)**.

The magnitude of the relevant neutron-nucleus interaction is characterized by a so-called coherent scattering cross-section σ (dimension m^2), which is related to a coherent scattering length b:

$$\sigma = 4\pi b^2 \qquad\qquad [6.32]$$

To a first approximation this scattering process may be regarded as a collision between two billiard balls, and the cross-section may be regarded as the effective area that the target nucleus presents to the incident beam of neutrons for the – essentially elastic – scattering process. The coherent scattering cross-section arises from interference effects between nuclei over relatively large distances. In other words, it reflects the structure factor $S(q)$, unlike the so-called incoherent scattering cross-section, which depends on uncorrelated motion of individual nuclei only. Similar to light scattering, coherent neutron scattering is peaked around $q = 0$, whereas the incoherent scattering contribution is approximately q-independent. It can be shown that in the q-range relevant for SANS the incoherent scattering contribution is usually orders of magnitude smaller than the coherent contribution.[17] However, to some extent this depends on the type of nuclei investigated, because the value of the scattering cross sections varies in an unpredictable manner from nucleus to nucleus. In particular, the hydrogen atom has a coherent scattering length $b_H = -0.374 \ 10^{-12}$ cm (b is in principle a complex number, the - sign here has no influence on the actual cross-section), and the deuterium atom has $b_D = 0.670$ 10^{-12} cm. On the other hand, the incoherent scattering length of hydrogen is an order of magnitude larger than that of deuterium, but still negligibly small when compared to the coherent scattering cross-section in the relevant q-range. The difference between the scattering lengths of hydrogen and deuterium is by far the largest amongst all nuclei, and it forms the basis for the so-called labeling method.

A marked difference in scattering contrast is obtained when polymers are synthesized with deuterium atoms rather than hydrogen atoms along the chain. In a polymer blend for instance, deuteration of one of the two components results in the relevant scattering contrast. The coherent scattering intensity $I(q)$ is given by:

$$I(q) = AK^2S(q) \qquad\qquad [6.33]$$

where A is an apparatus constant that can be determined independently by appropriate calibration of the instrument. The contrast factor K is a function of the difference in scattering lengths and the concentration of labeled (deuterated chains). It can be predicted independently as well. In this way, expressions for the free energy curvature similar to Eq. (6.21) (for light scattering) are obtained. Consequently, SANS and light scattering may be regarded as similar and useful techniques for determination of free energy curvatures or, more specifically, dilute solution properties such as interaction parameters and radii of gyration.

6.2.2 HEATS OF MIXING

At constant pressure, the heat released by a mixing process is proportional to the enthalpy of mixing ΔH_M. Such heats of mixing can be measured in very precise calorimeters.[1] In the case of simple disper-sive interactions, ΔH_M is positive, i.e., heat is consumed upon mixing. Specific interactions as well as compressibility effects cause negative contributions to ΔH_M, i.e., heat is released. Because the viscosities are too large, calorimetric measurements involving direct mixing of pure polymers have not been reported. One rather uses low-molecular weight analogues in the form of chemically similar oligomers. In case of polymer solutions, measurements of the heat of dilution (dilution of a stock solution with the solvent) proved a useful method to determine the enthalpy of mixing.[1]

6.2.3 GLASS TRANSITION TEMPERATURE

A practical tool, often used to determine whether a polymer mixture is miscible or not, is to compare the glass transition temperature(s) of the mixture with those of the constituents. Thanks to the spaghetti-type nature of polymer melts and the various types of chain entangle-ments involved, all polymers have glass transitions, where the rubbery melt is transformed into a hard, amorphous material upon passing through this transition (glass transition temperature, T_g) via cooling. In semi-crystalline materials, and even in highly crystalline materials in which crystallinity is never perfect, a glass transition exists below the crystalline melting point. On a microscopic scale, the actual transformation takes place in the amorphous material in between the crystalline spherulitic zones.[18] The precise nature of the glass transition in polymers is still subject to scientific debate,[19,20] details of which we feel are beyond the scope of this book. For all practical purposes, the glass transition in polymers is similar to a second order phase transition, as in supercooled liquids (in contrast to melting or crystallization, which are first order transitions). This implies that relevant thermodynamic variables that are related to second derivatives of the free energy, in particular the heat capacity C_p, the isothermal compressibility κ_T, and the coefficient of thermal expansion α, have a discontinuity at T_g. This discontinuity can be observed experimentally via Dynamic Mechanical Analysis (DMA) and Differential Scanning Calorimetry (DSC),[21] in general in the form of a peak at the T_g.

Polymeric mixtures that are thermodynamically miscible are miscible on a molecular scale. The miscible blend will exhibit a single glass transition temperature between the T_g's of the constituents with a sharpness of the transition similar to that of the respective constitu-ents. The practical suitability of the method depends on the extent to which the blend proves to be really thermodynamically miscible. A broadening of the transition or the merging together of the two individual transitions makes interpretation ambiguous. A good check on miscibility is that the (single) glass transition should show a systematic shift with varying composition. This relationship is not universally similar but has many variations. A typical empirical description often used is the Fox equation, which assumes composition averaged inversed additivity of the T_g's of the constituents:[22]

$$\frac{1}{T_g} = \frac{w_1}{T_{g1}} + \frac{w_2}{T_{g2}}$$ [6.34]

where w_i is the relevant composition fraction.

6.3 EXPERIMENTAL RESULTS AND MODEL VALIDATIONS

6.3.1 POLYMER SOLUTIONS

6.3.1.1 Polystyrene in cyclohexane

A well investigated model polymer solution is polystyrene (PS) in cyclohexane (CH). From an academic point of view, one would be inclined to state that systems in which the polymeric repeat unit and the solvent molecule are chemically similar, e.g., PS in toluene or ethyl benzene, should be investigated. This presumably allows assessment of thermodynamic effects originating mainly from entropy of mixing effects rather than energetic interactions. Problem however is that toluene is (indeed) a very good solvent for PS, so that phase separation phenomena are not easily observed in an accessible temperature range. An UCST has never been observed in a PS/toluene solution. In addition, specific interactions between still somewhat unsimilar chemical building blocks of PS and toluene cannot be ruled out a priori.

PS/CH apparently is the most simple model polymer solution with an experimentally accessible UCST and LCST demixing range, see Figure 6.1. It is likely that enthalpic interactions are of dispersive origin exclusively. It is widely accepted that this is a system of which the phase behavior should be predicted using entropy of mixing contributions, dispersive interactions and compressibility effects exclusively.

Depending on the purities of the respective components and the technique employed, a Θ-temperature of PS in CH around 30°C is reported.[23-26] With Small's group contribution scheme, solubility parameters of 9.1 and 8.2 $\sqrt{(cal/cc)}$ for PS and CH respectively, are predicted. This leads to a predicted Θ-temperature of 80K only, see also Section 4.2.3. Before we go into details of thermodynamic models investigated, it is interesting to discuss an experimental peculiarity of the Θ-temperature. First of all, the second virial coefficient A_2 is zero at this temperature. This has been confirmed with light scattering from dilute PS/CH solutions.[26] It is assumed that polymer chains

in a dilute solution apparently behave gaussian and ideal at the
Θ-temperature, due to compensation of self-avoiding expansion effects
of the isolated chain and shrinkage effects due to unfavorable dispersive interactions between polymer and solvent. Indeed, it was shown
o.a. by Chu *et al.*[26] that in dilute solutions of essentially monodisperse
PS in CH the radius of gyration scales with $\sqrt{M_w}$ at the Θ-temperature.
It was observed also that the coils appear to be in a swollen state above
and in a collapsed state below the Θ-temperature, with a diffuse
transition range in between, centered around the Θ-temperature. This
is again indicative for the balance of interactions at this temperature.
The actual temperature and molecular weight dependence of the
radius of gyration of the isolated chains could be described by master
curves appropriately normalized to Θ. Although FH-type models appeared to describe certain trends in the observed transition behavior
qualitatively, these master curves could not (yet) be predicted satisfactory.

Several attempts have been made to predict the complete phase
behavior of PS in CH, as determined with cloud point curves (CPCs),
spinodals, light scattering data etc. Saeki *et al.*[6] described the UCST
and LCST CPCs of PS/CH mixtures, as shown in Figure 6.1, using the
Patterson modified Flory-Pricogine EoS model (Section 4.3.2). Whilst
a value of $T^* = 7205$ K for PS[27] was used, values of T^* for CH (and
hence, the free volume difference parameter τ), the chain flexibility
parameter c and the cohesive energy density difference between PS
and CH were fitted to describe the observed phase behavior. With
these fit parameters, a temperature dependence of the interaction
parameter χ as defined in Eq. (4.64) was predicted. This prediction
could be tested against experimental UCST and LCST critical temperatures as a function of the chain length r. These critical temperatures correspond to a critical value of the interaction parameter:

$$\chi_c = \frac{1}{2}\left(1 + \frac{1}{r^{1/2}}\right)^2 \qquad\qquad [6.35]$$

Hence, one can chose to compare prediction and experiment on a χ-axis
or on a temperature axis. Saeki *et al.* presented a temperature axis,
where UCST and LCST critical temperatures were normalized to the
reduced temperature T^* of PS.[6] Results are shown in Figure 6.6.

The fitting procedure resulted in a consistent description of the
chain length dependence of both UCST and LCST critical temperatures. Such description was obtained with the fit parameters shown
in Table 6.1. The value found for the chain flexibility parameter c
represents that of a fully flexible chain. This could bare relevance to
the fact that the phase diagram and the relevant temperature range

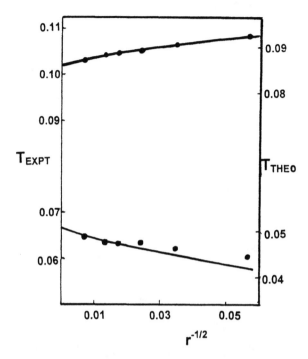

Figure 6.6. Comparison of experimental UCST and LCST (reduced) critical temperatures with the Flory-Pricogine EoS prediction. (Adapted, by permission from S. Saeki, N. Kuwahara, S. Konno, and M. Kaneko, *Macromolecules*, **6**, 246 (1973)).

Table 6.1: EoS parameters of PS solutions.

Mixture	T^*, K	c	τ^2	$10^3 \, v^2$
PS-CH	4720	1.01	0.119	15.9
PS-MCH	4870	1.14	0.105	15.8
PS-Toluene	4979	1.57	0.095	0^a

a: assumed values

investigated is relatively close to the Θ state and hence, the chain would indeed behave quasi ideal. On the other hand, it is incompatible with the observation as such that phase separation occurs and that the isolated chains swell and shrink around the Θ-condition. The latter effect was not incorporated in the EoS model as such. The value of the characteristic dispersive energy parameter v is related to the solubility parameter difference between PS and CH:

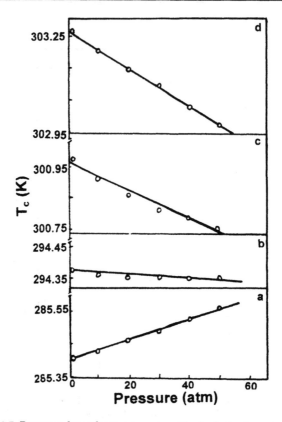

Figure 6.7. Pressure dependence on upper critical solution temperature in PS/CH mixtures for the samples: (a) $M_w = 3.7 \times 10^4$; (b) $M_w = 11 \times 10^4$; (c) $M_w = 145 \times 10^4$. (Adapted, by permission from, S. Saeki, N. Kuwahara, M. Nakata, and M. Kaneko, *Polymer*, **16**, 445 (1975)).

$$v^2 = \frac{(\delta_{PS} - \delta_{CH})^2}{\delta_{CH}^2} \qquad\qquad [6.36]$$

The value found for v corresponds to a solubility parameter difference of 1.03 $\sqrt{(cal/cc)}$, close to the predicted difference of 0.9 $\sqrt{(cal/cc)}$ using Small's scheme. The value found for the characteristic free volume difference parameter τ^2 would translate into a typical difference in thermal expansion coefficient between PS and CH of about $5.10^{-4}\ K^{-1}$, which is expected for typical polymer solutions, see also Table 4.1.

Saeki *et al.* also investigated the pressure dependence of the UCST of PS/CH over the pressure range 1 to 50 bar.[28] The changes are small, but CPCs could be detected with sufficient precision with the aid of a He-Ne laser. Results for various molecular weights are shown in Figures 6.7. Interestingly, with increasing pressure, UCSTs tend to shift to higher temperatures at the lowest molecular weight

Table 6.2: Pressure dependence of UCST temperatures in PS/CH.

M_w, 10^4 g/mol	UCST, 1 bar (K)	$(dT/dp)_c$, 10^{-3} K/bar	
		Experimental	Flory
3.7	285.41	3.14	-25.6
11.0	294.38	-0.52	-31.8
67.0	300.97	-4.40	-34.1
145.0	303.26	-5.64	-35.3

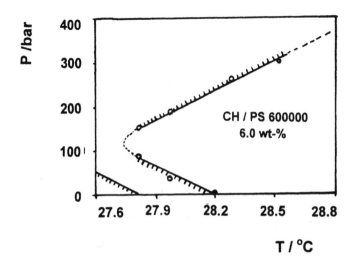

Figure 6.8. Demixing curve of a solution of 6% PS in CH; the pressure-temperature region of the two phase system is indicated by hatching. The filled circle gives the cloud point curve of the solution at rest. The parallel line shifted towards lower temperatures is taken from Saeki et al.[28] (Adapted, by permission, from B.A. Wolf and H. Gaerisen, *Coll. & Polym. Sci.*, **259**, 1214 (1981)).

and to lower temperatures at higher molecular weights. The magnitude of the pressure dependence increases with increasing molecular weight.

Flory-Pricogine EoS predictions of the pressure dependence of UCST critical temperatures are shown in Table 6.2. The theory shows an UCST decrease with increasing pressure. The effect increases with molecular weight. However, the size is one order of magnitude too large and the UCST increase at low molecular weight is not predicted.

Wolf and Geerissen[29] measured the pressure dependence of the UCST of a monodisperse PS (M=600.000) in CH solution at 6% weight fraction of PS in the range 1-300 bar. At low pressures, a drop in UCST

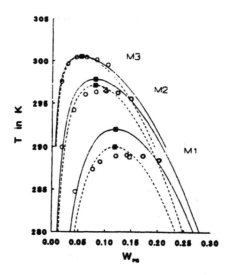

Figure 6.9. Experimental spinodal[31] and critical[32] data for the systems PS, $M_{1w} = 51,000$; $M_{2w} = 166,000$; $M_{3w} = 520,000$ in CH. Predicted critical points (solid squares) and spinodal curves according to S&S (dotted line) and HH (solid line) models. (Adapted, by permission, from A. Stroeks and E. Nies, *Macromolecules*, **23**, 4092 (1990)).

was observed as well. However, the corresponding increase in solubility appeared to be optimal around 120 bar: at higher pressures the UCST increased again, such that at 300 bar a decrease in solubility was obtained when compared to atmospheric conditions, see Figure 6.8.

A sizable amount of spinodal data in the UCST range is available. These were determined using a special light scattering technique, the so-called Pulse Induced Critical Scattering (PICS).[30,31] The divergence of scattered intensity in the metastable region of the phase diagram is monitored via rapid and repetitive quenching of a thin capillary filled and sealed with polymer solution. Normally, a simple plot of the inverse of the scattered intensity versus quench temperature allows determination of spinodal temperatures via extrapolation. Thermodynamic models can be fitted relatively easy to spinodals. Fitting binodals over an extended concentration range is more difficult because it involves numeric solution of non-linear equations for the chemical potentials. Experimental UCST spinodal[31] and critical[32] data for monodisperse PS/CH solutions of indicated molecular weight are shown in Figure 6.9.

Binodal and also spinodal data were used by Nies *at al.*[40,41] to describe the phase behavior of PS/CH solutions. This was done using the most general form of the EoS method, in which both cell free volume and lattice vacancies are allowed for (Section 4.3.3). Both the Simha-Somcynsky (S&S) and the Holes and Huggins (HH) models were tested. The most important features of these models can be summarized as follows: For the calculation of the combinatorial entropy of mixing of holes and segments and of different segments S&S use the original FH assumptions. In the HH model this combinatorial term is modified to include the Huggins correction. Hence, a lattice coordination number z enters explicitly into the HH model.

For the calculation of the energetic contributions to the free energy of mixing both models assume Lennard-Jones type interactions. Hence, such interactions are characterized in the relevant free energy expressions with the maximum attraction energy parameter ε^* and the segmental repulsive volume v^*. In the corresponding cell partition function the occupied lattice site fraction is the most important structure parameter. In the HH model a more detailed structure parameter is used, namely the intersegmental contact fraction. It is a topologically more precise parameter and was already introduced by Huggins.[36] Depending on the lattice coordination number and the actual site fraction, the contact fraction is a measure for the number of intersegmental contacts the chain segments can make.

It was shown that the modifications of the S&S model produce relatively minor effects on the EoS of the pure components,[33] but lead to improved predictions of the miscibility behavior, especially with regard to pressure dependence.[34] Since the lattice coordination number explicitly enters into the relevant partition functions in the HH model, the value of this parameter was fixed in order not to introduce yet (another) adjustable parameter. A value of z=12 was chosen, corresponding to a closest packing of spheres. Together with the free volume degrees of freedom, this may represent a more or less a disordered liquid.

To confront theory with experiment the molecular parameters in the models were fitted. The pure component parameters ε^*, v^* and c were determined from experimental EoS data.[36,37] Only two fit parameters are needed for the mixture, namely $\varepsilon^*_{PS/CH}$ and $v^*_{PS/CH}$. These two parameters were obtained by fitting the UCST critical point of the high molecular weight M3 mixture in Figure 6.9. The complete set of parameters thus obtained is presented in Table 6.3. One cannot conclude very much from these parameters as such, because it are basically fit parameters. Nevertheless, the fact that for both models values of both mixing parameters are in between those of the pure components and do not change significantly upon mixing may be physically sound if one considers the simple dispersive interactions involved. On the other hand, a flexibility parameter value of 1.8 needed to fit the EoS of CH seems in contrast with the fact that PS is

Table 6.3. Molecular parameters for PS and CH according to the S&S
and HH models (from Stroeks and Nies[34]).

		ε^*, J/mol	v^*, 10^{-5} m^3/mol	c
PS				
	S&S	6968.8	9.95	0.77
	HH	6690.9	9.84	0.85
CH		6543.3	9.36	1.80
PS/CH				
	S&S	6727.8	9.71	
	HH	6559.5	9.67	

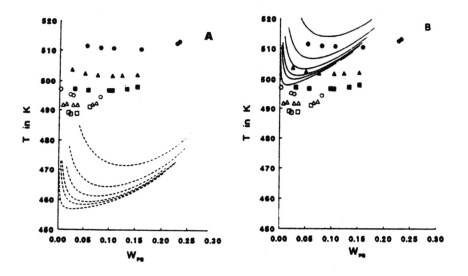

Figure 6.10. Experimental LCST cloud point data (from Figure 6.1) and
predicted spinodal curves according to the (A) S&S and (B) HH models.
(Adapted, by permission, from A. Stroeks and E. Nies, *Macromolecules*, **23**,
4092 (1990)).

considered a monomer in both models. Moreover, in the original theory
of Flory a theoretical value of c larger than 1 is not expected.

Note that differences between HH and S&S remain relatively
small with regard to the UCST phase behavior, see Figure 6.9. For
both models the calculated change of critical temperature and compo-

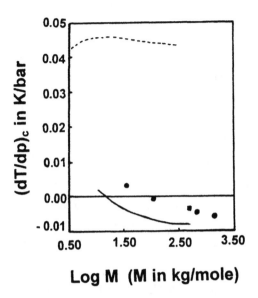

Log M (M in kg/mole)

Figure 6.11. Pressure coefficient $(dT/dP)_c$ versus the logarithm of the molar mass. Experimental data of Saeki et al.[28] (circles) and Wolf and Geerissen[29] (squares). Predictions according to S&S (dotted line) and HH (solid line) models. (Adapted, by permission, from A. Stroeks and E. Nies, *Macromolecules*, **23**, 4092 (1990)).

sition with molar mass is too small. The curvature of the calculated spinodals appears to be somewhat too big. With the same set of molecular parameters, LCST spinodals were predicted as well. These spinodals were compared with the LCST binodals as measured by Saeki *et al.* (Figure 6.1.). Results are presented in Figure 6.10. Predicted HH spinodals are more close to the real critical points than the S&S spinodals. Also here the predicted change of critical coordinates is somewhat too small. *Grosso modi*, the agreement between theory and experiment is perhaps not bad, if it is realized that LCST demixing is predicted from a fit to one UCST critical point some 180K lower.

Differences between HH and S&S models become more pronounced when the pressure dependence of the phase behavior is predicted, using again the same molecular parameters fitted to the UCST critical point. In Figure 6.11, predictions of the initial slope dT_c/dp of the critical temperature is compared with the experimental data of Saeki *et al.* (Figure 6.7) and the data of Wolf and Geerissen (Figure 6.8). The S&S model predicts the slope an order of magnitude too large and positive. The HH model does predict the change of sign versus molecular weight. Computed and measured values of the slopes differ by approximately a constant amount irrespective of

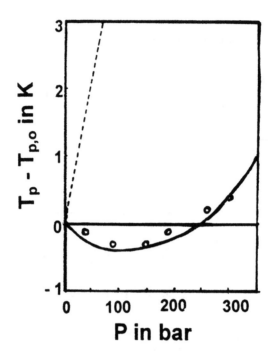

Figure 6.12. Difference between critical temperatures at elevated and atmospheric pressure versus pressure. Predictions according to S&S (dotted line) and HH (solid line) models. (Adapted, by permission, from A. Stroeks and E. Nies, *Macromolecules*, **23**, 4092 (1990)).

molecular weight, but the molar mass dependence is predicted quite well by the HH model.

Model predictions of the pressure dependence of the UCST critical point as such (at constant molecular weight) were also compared with the data of Wolf and Gaerisen. Results are shown in Figure 6.12. The S&S model predicts a simple increase with pressure, and the slope is about 10 times too large and has wrong sign. The HH model gives a correct prediction of the observed effect, i.e., the initial slope, the maximum UCST drop around 120 bar and a decrease in miscibility above this pressure.

Nies et al.[34] also predicted the so-called negative excess volumes upon mixing, i.e., the (expected) volume contraction due to the free volume effects. It was also suggested that the peculiar behavior of the pressure dependence of the UCST critical point is correlated to a change in curvature in the concentration dependence of this excess volume.[34] This needs however verification. In any case, the HH model appears to be capable of predicting relatively subtle pressure effects and sets it apart from other EoS models.

6.3.1.2 Other polymer solutions

Saeki *et al.*[6] also measured CPCs of two other PS mixtures, namely in methylcyclohexane (MCH) and toluene. PS in MCH shows UCST and LCST behavior and the corresponding critical points are similar to those of the PS/CH solution. Hence, the corresponding molecular parameters as predicted by the Patterson EoS model are also similar, see Table 6.1. In the toluene solution an LCST is observed only, some 60K higher than in the (M)CH solutions. Indeed, the solubility parameter difference appears to be relatively small so that an UCST is lacking (or occurs at subzero temperatures close to or even below the vitrification temperatures). Assuming a characteristic dispersive interaction parameter of zero (Table 6.1), the LCST demixing can be predicted which results in a characteristic free volume difference parameter somewhat smaller than in (M)CH solutions, see Table 6.1. This is consistent with the higher LCST and corresponding increased miscibility. Note that the flexibility parameter c appears to vary in an impredictible manner from solution to solution.

In mixtures where miscibility is rather poor, UCST and LCST may merge together to form and hour-glass shaped phase diagram. In Figure 6.13 the phase diagrams of mixtures of PS in acetone[38] are shown. At relatively low molecular weights of PS separate UCSTs and LCSTs can be observed but merge together until at a molecular weight around 20,000 an hour-glass shaped phase diagram is obtained.

Sanchez and Lacombe used the Lattice Fluid (LF) theory to describe the phase behavior of several polyisobutylene (PIB) solutions.[39] These solutions were chosen because, according to the LF theory (Section 4.3.3), a polymer solution in equilibrium with its own vapor is capable of reaching an LCST critical point prior to the liquid-vapor critical point T_c. For several PIB/solvent mixtures a range $0.7 < LCST/T_c < 0.9$ was observed.[40]

Pure component EoS parameters were obtained by fitting vapor pressure data of the solvents (the complete series pentane to decane, cyclohexane and benzene) and liquid density data of PIB. Heats of mixing at infinite dilution $\Delta H_m(\infty)$ were used to determine the energetic interaction parameter χ (Eq. (4.76)) between PIB and the seven hydrocarbon solvents. The energetic interaction parameter could be expressed o.a. in two equivalent forms χ and ΔP^*. They are shown in Table 6.4. The parameter ΔP^* represents the change in cohesive energy density upon mixing at the absolute zero of temperature. As expected for the non-polar solutions, the calculated values are all positive. Thus, at absolute zero the heats of mixing would all be positive. However, only PIB/benzene has a positive $\Delta H_m(\infty)$ at 298K, whilst the heat of mixing of all other solutions is negative. In terms of the LF theory, such negative heats appear to be caused by the

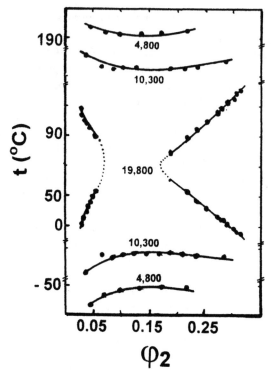

Figure 6.13. Phase diagram of the PS-acetone mixture for fractions of indicated molecular weight showing UCST and LCST for low molecular weights and the "hour-glass" CPC for the 19,800 fraction. (Adapted, by permission, from K.S. Siow, G. Delmas, and D. Patterson, *Macromolecules*, **5**, 29 (1972)).

Table 6.4. Interaction parameters and experimental and theoretical volumes of mixing and LCST for polyisobutylene solutions (from Sanchez and Lacombe[39]).

	$\Delta H(\infty)$ J/mol	χ 10^{-2}	ΔP^* MJ/m^3	$\Delta V_m/V_0$, 10^{-2}		LCST, K	
				exp.	calc.	exp.	calc.
pentane	-201	4.98	10.4	-1.27	-1.83	344	<298
hexane	-159	3.24	6.04	-0.86	-1.25	402	<298
heptane	-100	2.59	4.91	-0.62	-0.92	442	303
octane	-67	2.13	3.89	-0.48	-0.73	477	375
decane	-31	1.44	2.46	-0.29	-0.48	535	470
cyclohexane	-38	2.44	5.61	-0.14	-0.44	516	440
benzene	1090	8.18	20.7	0.34	0.20	534	487

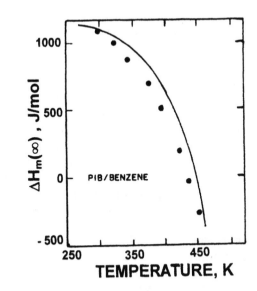

Figure 6.14. Comparison of experimental (circles) and theoretical (line) heats of mixing fir dilute PIB/benzene solutions. (Adapted, by permission, from I.C. Sanchez and R.H. Lacombe, *Macromolecules*, **11**, 1145 (1978)).

tendency of the solvent to contract when a small amount of polymer is added. The magnitude of the contraction is proportional to the compressibility of the solvent. It is an energetically favored process, compensating against the unfavored energetic cross-interaction term χ, because it results in more intermolecular interactions of lower potential energy amongst the solvent molecules exclusively. On the other hand, such contraction also has its draw-back on (unfavorable) entropic contributions to the free energy of mixing, which will lead to LCST demixing.

Solutions with a negative room-temperature heat of mixing also show negative volume changes, whilst the PIB/benzene solution shows both a positive heat of mixing and a volume expansion. The values of ΔV_m are the maximum observed volume changes at the 50/50 composition. The calculated values have the correct sign but the quantitative agreement is not very good.

Although benzene has a large and positive heat of mixing with PIB at room temperature, a decrease with increasing temperature was found. Heats of mixing also become negative above 435K, see Figure 6.14. The predicted curve from the LF model was calculated using the pure component EoS parameter and the room temperature interaction parameter determined from $\Delta H_m(\infty)$. Apparently, compared to the other mixtures, compressibility effects in PIB/benzene become important at somewhat higher temperatures.

LCST critical temperatures were calculated as well and compared with experiment, see Table 6.4. Again, pure component parameters and room-temperature heats of mixing data were used exclusively. For pentane and hexane the calculated value of the effective FH interaction parameter (a free energy parameter in the LF model) was greater than 0.5 and LF theory incorrectly predicted immiscibility of these two systems at 298K. For the remaining solvents the calculated LCSTs are substantially lower than observed. Hence, from this relatively severe mismatch (e.g., when compared to predictions of the HH model) it may be concluded that entropic contributions to the free energy of mixing, that are associated with the compressibility effect, are not adequately incorporated in the LF model.

Since the light scattering technique is in particular suited to study the thermodynamics of dilute polymer solutions via Zimm analysis, a relatively large amount of data on virial coefficients and FH interaction parameters is available.[24,41-43] A large number of experimental FH interaction parameters of dilute polymer solutions were collected by the present authors and analyzed in terms of temperature and molecular weight dependence.[44] In an attempt to find explanations behind the systematic underestimation of interaction parameters (see Section 4.2), it was assumed in analogy that the interaction parameter χ could be written as a sum of an enthalpic contribution χ_h and an entropic contribution χ_s:

$$\chi = \chi_s + \frac{\chi_h}{T} \qquad\qquad [6.37]$$

Solutions investigated included PS/CH, PS/toluene, PS/decalin, PS/MEK, PMMA/butyl chloride and PS/benzene. Experimental χ-parameters covered a range between 0.43 and 0.52 around the Θ-temperatures. From the experimental data a strong correlation between χ_s and χ_h values was observed, see Figure 6.15. Higher χ_s values correspond to lower χ_h values in a seemingly linear manner. In fact all data, except for PS/toluene, fall on the same line.

The strong correlation is partly trivial because all solutions were investigated around their respective Θ-temperatures, in all cases approximately around ambient. Therefore, in all cases, Eq. (6.37) adds-up to a total, almost constant χ value of 0.5 so that χ_s and χ_h must be correlated in the (limited) temperature region investigated. The question "Why do these totally different systems have almost identical

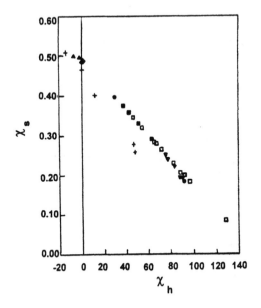

Figure 6.15. Entropic parameter χ_s versus enthalpic parameter χ_h. (Adapted, by permission, from M.A. van Dijk and A. Wakker, *Polymer,* **34,** 132 (1993)).

Θ-temperatures" is thus identical to "Why are the χ_s and χ_h parameters linearly related".

Another interesting observation can be made on the molecular weight dependence of the χ_s and χ_h of a given solution. Figure 6.16 shows the molecular weight dependence of χ_s of three different solutions. A steep increase with molecular weight at low values and a leveling off at higher values is observed. Comparing these results with Figure 6.15, it can be concluded that with increasing molecular weight, the χ_s increases and the χ_h decreases in such a way that both are linearly related.

It is tempting to conclude that there is a universal type of behavior of the interaction parameter in dilute polymer solutions. The value of χ_s increases with molecular weight and apparently also to a sort of limiting value at infinite molecular weight that depends on the particular polymer- solvent mixture. The resemblance with the Huggins correction, Eq. (3.41), is striking. Unfortunately there is as yet no clear a correlation between molecular structure and the lattice coordination number z.

An alternative to light scattering from dilute polymer solutions is to determine directly the free energy curvature from the measured

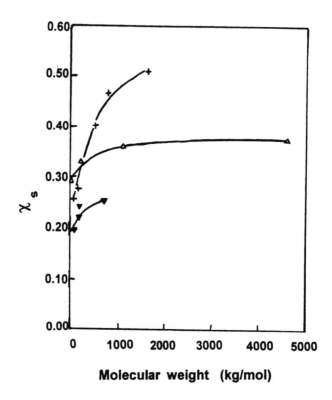

Figure 6.16. Molecular weight dependence of the entropic parameter χ_s of three different polymer solutions. Curves serve as a guide to the eye. (Adapted, by permission, from M.A. van Dijk and A. Wakker, *Polymer*, **34**, 132 (1993)).

Rayleigh ratios, see Eq. (6.21). Advantage is that this can be done irrespective of concentration. The present authors investigated the phase behavior of PS in methyl acetate in this way.[45] Free energy curvatures were determined in the homogeneous one-phase region, in between the UCST and LCST of this mixture, over a broad range of temperatures (300-400 K) and concentrations (0.5%-10.3%). Results are shown in Figure 6.17. Cloud point curves were determined as well, and are shown in Figure 6.18.

The free energy curvature is smallest for sample III (3.25 wt%) and it is also nearly zero at both the high- and low-temperature side. This indicates that the concentration of sample III is very close to critical, which is also in agreement with the CPCs shown in Figure

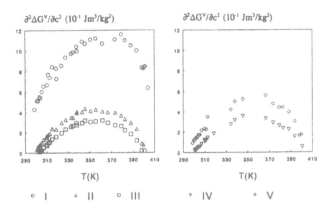

Figure 6.17. Values of the free energy curvatures for samples I (0.5 wt%), II (1.6 wt%), III (3.25 wt%), IV (6.5 wt%), and V (10.3 wt%). (Adapted, by permission, from A. Wakker, F. van Dijk, and M.A. van Dijk, *Macromolecules*, **26**, 5088 (1993)).

6.18. Extrapolation towards zero curvature results in a UCST critical temperature of 301.5K and an LCST critical point of 402.5K. Free energy curvature and hence, thermodynamic miscibility, increases both at lower and at higher concentrations. Optimum miscibility is also observed approximately halfway between UCST and LCST. The complete dataset is suitable to test model predictions, but no truly predictive theory could as yet thus be tested.

6.3.2 POLYMER BLENDS

6.3.2.1 Polystyrene in polyvinyl methyl ether

A useful model mixture often used to study the thermodynamics of high molecular weight polymer mixtures (where immiscibility is the rule, and miscibility the exception) is PS in polyvinyl methyl ether (PVME; $(-CH_2-CHOCH_3-)_n$). Both polymers are fully amorphous and have an experimentally accessible range of miscibility in between the (single) glass transition temperature and the LCST. An UCST was never observed.

The neutron labeling technique proved to be useful to study the phase behavior of PS/PVME mixtures using SANS. Deuterated PS (PSD) and PVME can be relatively easy made (and molecular weights controlled) on a laboratory scale. Excellent SANS studies on PSD/PVME mixtures have been reported.[46-48] Of course, binodals of this mixture could be determined easily as well on thin samples using hot stage microscopy or simple laser light attenuation. Actually, CPCs

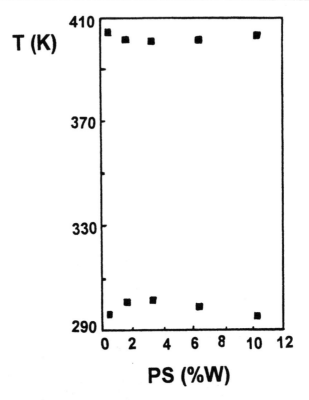

Figure 6.18. CPCs of PS in methyl acetate. (Adapted, by permission, from A. Wakker, F. van Dijk, and M.A. van Dijk, *Macromolecules*, **26**, 5088 (1993)).

of PS/PVME mixtures were already determined by McMaster[49] and were also reported by Han[46] on the deuterated mixture.

In the basic understanding of thermodynamics as well as dynamics of phase-separation, PS/PVME proved a useful model system because dynamics of phase separation could be studied as well by means of quenching-type light scattering experiments.[10,46,50] The main results of SANS and light scattering experiments on PS(D)/PVME mixtures are presented below.

A typical phase diagram of PSD in PVME is shown in Figure 6.19. CPCs were determined from light scattering and spinodals were determined from extrapolation of SANS data. Note that the LCST is skewed towards the higher molecular weight PSD-poor side of the phase diagram.

Similar to the quantitative analysis of light scattering data, the relevant free energy curvatures can in principle be extracted from SANS intensities extrapolated to zero scattering angle. However, it has been shown, o.a. by de Gennes[7] that a more straightforward

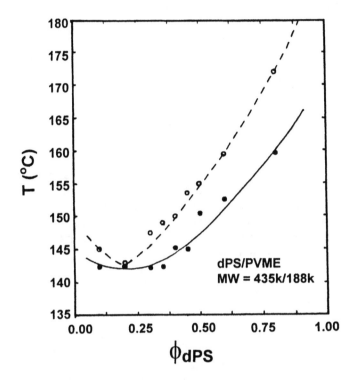

Figure 6.19. Phase diagram of PSD/PVME with molecular weights 4.35×10⁵/1.88×10⁵. Solid and dotted lines are cpc and spinodal, respectively. (Adapted, by permission, from C.C. Han, M. Okada, Y. Muroga, B.J. Bauer, and Q. Tran-Cong, *Polym. Eng. Sci.*, **26**, 1208 (1986)).

procedure exists to obtain the relevant binary interaction parameter χ. This procedure, also known as the Random Phase Approximation,[51] is based on the principle that in polymer melts the coils have a gaussian and ideal conformation so that the structure factor S(q) can be expressed in terms of the single chain form factors of the respective constituents plus a binary interaction term in which the interaction parameter χ appears. This interaction parameter is equivalent to the interaction parameter as defined in the original Flory-Huggins model, provided that it is given an (arbitrary) concentration dependence. The relevant expression for the structure factor S(q) reads:

$$\frac{k_N}{S(q)} = \frac{1}{\phi_A r_A v_A S_D(A)} + \frac{1}{\phi_B r_B v_B S_D(B)} - \frac{2\chi}{v_0} \qquad [6.38]$$

Table 6.5. Molecular weight and polydispersity of three series of PSD/PVME mixtures (from Han et al.[48]).

	PSD		PVME	
	M_w	M_w/M_n	M_w	M_w/M_n
L series	230,000	1.14	389,000	1.25
M series	402,000	1.42	210,000	1.32
H series	593,000	1.48	1,100,000	1.26

where k_N is a (coherent) scattering contrast factor, ϕ_i, r_i, v_i are the volume fraction, chain length, and molar volume of component i; $S_D(i)$ is the single chain form factor (Debye function) of component i, and v_o is the molar volume of a reference segment (to be taken equal to that of one of the components or a part of it). The single chain form factor $S_D(q)$ may be represented analytically:

$$S_D(q) = \frac{6}{q^2 R_g^2}\left(\exp\left(\frac{-q^2 R_g^2}{3}\right) - 1 + \frac{q^2 R_g^2}{3}\right) \qquad [6.39]$$

where R_g is the radius of gyration of the coils of constituent i. It can be shown that by expansion of the Debye function in powers not higher than q^2 the Ornstein-Zernike form (Eq. (6.24)) is retained. In that case, the relevant correlation length ξ has become a function of the interaction parameter.

Hence, by fitting the experimentally determined structure factors $S(q)$ of PSD/PVME mixtures to Eq. (6.38), the interaction parameter χ can be obtained as a function of mixture composition and temperature. Correlation lengths ξ can be obtained as well. Spinodal temperatures of the blend can be obtained at various compositions equivalently by extrapolation of ξ^{-2} or $S(0)^{-1}$ versus temperature.

Han et al. thus studied three series of mixtures with varying molecular weights, polydispersities, and skewness, see Table 6.5. In Figure 6.20 the scattering structure factor $S(q)$ is shown versus q for the 30/70 sample of the M series at various temperatures. The full curves are best fits according to Eq. (6.38). Note the strong increase of scattering intensity $S(q=0)$ and the decrease of the width of $S(q)$ (increase of the correlation length ξ) with increasing temperature (and decreasing distance from LCST spinodal temperature).

In Figure 6.21, reciprocal square correlation lengths ξ^{-2} are shown versus temperature for various compositions of the H series results. The ξ^{-2} values can be represented by a straight line near the spinodal temperature which is at $\xi^{-2}=0$. A similar plot of $S(q=0)^{-1}$

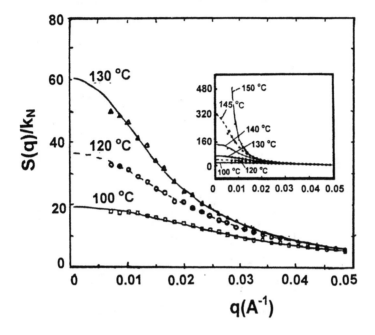

Figure 6.20. Scattering intensity S(q) from SANS experiment for M series PSD/PVME sample of 30 wt% PSD versus q for 100, 120, and 130°C. Experiments with several other temperatures are displayed at reduced scale in the insert. (Adapted, by permission, from C.C. Han, B.J. Bauer, J.C. Clark, Y. Muroga, Y. Matsushita, M. Okada, Q. Tran-Cong, T. Chang, and I. Sanchez, *Polymer*, **29**, 2002 (1988)).

versus temperature is shown in Figure 6.22. Again, a good linear relation is obtained near the spinodal. Spinodal temperatures could thus be obtained consistently from the intercepts of both ξ^{-2} and $S(q=0)^{-1}$ going to zero. A complete dataset of spinodal temperatures as a function of molecular weights and composition can be found in ref. 48, Appendix 2.

In Figure 6.23, χ/v_0 values are plotted versus 1/T for the H series sample at various compositions as indicated in the figure. Within the (limited) temperature range investigated the interaction parameters are negative and follow a 1/T - dependence. The interaction parameters strongly depend on composition: the higher the PSD concentration, the more negative the interaction parameters become at a given temperature. Such concentration dependence is shown again in Figure 6.24 for all three series. It seems that χ/v_0 varies linearly with composition except that there may be a very small curvature at

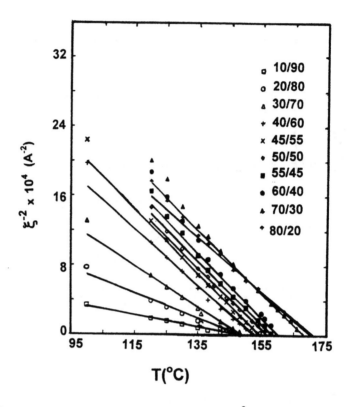

Figure 6.21. Reciprocal square correlation length ξ^{-2} plotted versus T for various compositions for H series sample. (Adapted, by permission, from C.C. Han, B.J. Bauer, J.C. Clark, Y. Muroga, Y. Matsushita, M. Okada, Q. Tran-Cong, T. Chang, and I. Sanchez, *Polymer*, **29**, 2002 (1988)).

$\phi_{PSD}<0.3$. Note that the interaction parameters hardly depend on the molecular weight of the mixture.

Before we will discuss the above results in terms of thermodynamic models that were fitted to it, it is useful to mention the results that Hashimoto *et al.*[10] obtained from phase separation kinetics studies via spinodal decomposition (SD) of a 30/70 PS/PVME blend. In Figure 6.4, an example is given of the growing light scattering intensity versus time after quenching of a thin PS/PVME film into the unstable region (above the LCST spinodal, quench depth 1.8 K). As predicted by the theory of SD, a distinct scattering maximum appeared with an exponential growth rate with time proportional to $R(q_{max})$, see Eq. (6.7). The position of this peak did not change during a definite time span (in this experiment typically half to one hour), depending on the quench depth, i.e., the distance from actual quench temperature to the spinodal temperature. In the later stage of SD, the

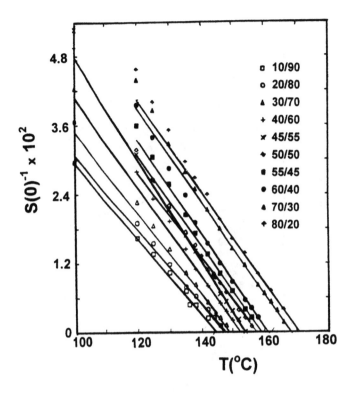

Figure 6.22. $S(0)^{-1}$ plotted versus T for H series sample. (Adapted, by permission, from C.C. Han, B.J. Bauer, J.C. Clark, Y. Muroga, Y. Matsushita, M. Okada, Q. Tran-Cong, T. Chang, and I. Sanchez, *Polymer*, **29**, 2002 (1988)).

intensity increase with time started to deviate from exponential behavior and the scattering maximum shifted towards smaller q values, corresponding to the onset of macroscopic phase separation in the form of coarsening of the phase separating domains. The higher the superheating, the earlier the stage where this coarsening started.

In the early stage of SD, the relaxation rate R(q) could thus be determined from the exponential increase of the scattered intensity. Once these relaxation rates were derived, "apparent" diffusion coefficients D_{app} could be determined from the intercepts of plots of $R(q)/q^2$ versus q^2, see also Eq. (6.3). Results are shown in Figure 6.25 (a). A linear relation is obtained for the various quenching temperatures, indicative of the suitability of SD theory in the early stage. Figure 6.25 (b) shows a plot of D_{app} as a function of temperature. The larger the quench depth, the faster the diffusion, i.e. the faster the kinetics of SD

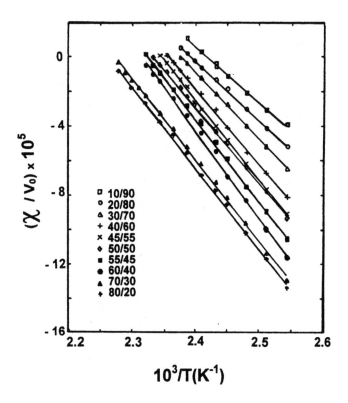

Figure 6.23. Reciprocal temperature dependence of χ/v_0 for various compositions of the H series sample. (Adapted, by permission, from C.C. Han, B.J. Bauer, J.C. Clark, Y. Muroga, Y. Matsushita, M. Okada, Q. Tran-Cong, T. Chang, and I. Sanchez, *Polymer*, **29**, 2002 (1988)).

(and, ultimately, the phase separation). The intercept $D_{app} = 0$ resulted in an extrapolated spinodal temperature of 99.2°C at this particular composition.

The negative interaction parameters observed in the homogeneous one-phase temperature region below the LCST, as extracted from from the SANS experiments of Han *et al.* indicate that specific interactions definitely play a role in the phase behavior of PS(D)/PVME mixtures. Compressibility can however not be ruled out a priori, although its influence on the phase behavior is expected to be significantly smaller than in polymer solutions. In any case, solubility parameters of PS (9.1 √(cal/cc)) and PVME (8.5 √(cal/cc)) appear to be too far apart to bring about miscibility in the high molecular mixtures.

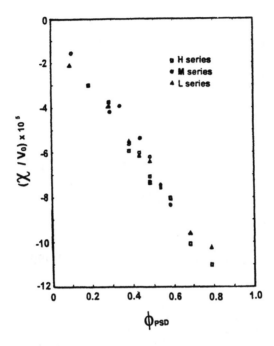

Figure 6.24. Interaction parameter χ/v_0 for all three series of samples at 130°C. (Adapted, by permission, from C.C. Han, B.J. Bauer, J.C. Clark, Y. Muroga, Y. Matsushita, M. Okada, Q. Tran-Cong, T. Chang, and I. Sanchez, *Polymer*, **29**, 2002 (1988)).

CPCs of PS/PVME mixtures were unsuccessfully predicted by McMaster using the Flory EoS theory.[49] McMaster came to the conclusion that specific interactions should be included. It is therefore instructive to predict the temperature dependence of the (negative) interaction parameters using ten Brinke's model[52] (Section 4.4.1), which does take into account specific (and dispersive) interactions, but not the compressibility of the mixture. This we have done for the interaction parameters determined by Han *et al.* on a 50/50 PSD/PVME mixture, see Figure 6.26. The repeat unit of PVME was chosen as the unit lattice cell ($v_0 = 53$ cm^3/mol). Using the solubility parameter difference $\Delta\delta = 0.6$ √(cal/cc), a dispersive interaction energy $U_2 = 0.0191$ kcal/mol is obtained. Using this value as an input parameter, the ten Brinke model was fitted to the data points. From this fit, a specific interaction energy $U_1 = -0.24$ kcal/mol, a degrees of freedom parameter q = 14.7 and a lattice coordination number z = 14.8 were obtained. The latter parameters should be regarded as phenomenological. The (small) value U_1 could be regarded as characteristic

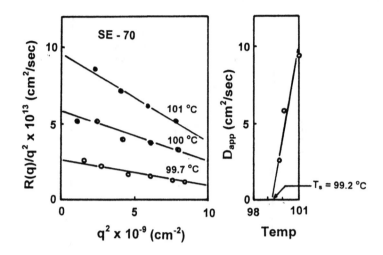

Figure 6.25. (a) $R(q)/q^2$ versus q^2 at three different isothermal phase separation temperatures, the arrows indicating the position of q_{max}^2; (b) the temperature dependence of D_{app} from which the spinodal temperature $T_s=99.2°C$ is deduced. (Adapted, by permission, from T. Hashimoto, J. Kumaki, and H. Kawai, *Macromolecules*, **16**, 641 (1983)).

for a weak dipole interaction, sufficient to compensate for the solubility parameter difference and hence to bring about miscibility, see also Section 4.4.2.

The possibility of a combined effect of specific interactions and compressibility on the LCST phase behavior was investigated by Sanchez and Balasz using their LF model modified to allow for specific interactions.[54] The incompressible version of this model is essentially ten Brinke's model. Spinodal data extracted from the SANS data of Han were compared with model predictions. Pure component EoS parameters were taken from the literature. In Figure 6.27 (a), the experimental spinodal of the L-series sample is shown together with the predicted spinodal of the LF model in which the specific interaction energy U_1 was set to zero. The calculated LCST critical temperature was matched to the experimental one only by adjusting a non-specific interaction energy parameter ε_{12}. In the context of the LF model, the heat of mixing $\Delta\varepsilon$ is given by:

$$\Delta\varepsilon = \varepsilon_{11} + \varepsilon_{22} - 2(\varepsilon_{12} + U_1) \qquad [6.40]$$

In this particular case, where $U_1 = 0$, the value of ε_{12} was such that the heat of mixing was slightly negative. Hence, compressibility

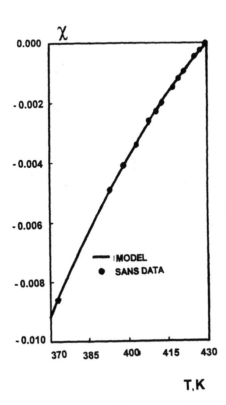

Figure 6.26. The χ as a function of temperature in PS/PVME according to the ten Brinke model. (Adapted, by permission, A. Wakker and M.A. van Dijk, *Polym. Networks Blends*, **2**, 123 (1992)).

caused miscibility at lower temperatures and LCST demixing (because of the entropy penalty associated with compressibility) at higher temperatures. Interestingly, the calculated critical composition ϕ_c appears to be independent of the choice of ε_{12} and is correctly predicted. In the classical theory, ϕ_c is rich in the component with the smallest chain length, in this case PS. However, the observed ϕ_c is rich in PVME, correctly predicted by the LF model. The LF model appears to qualitatively predict that due to compressibility the ϕ_c for an LCST is rich in the component which has the smallest pure component ε_{ii}, which is in this case PVME.

The actual predicted spinodal is too narrow. It was shown that the spinodal can be broadened by "switching on" the specific interactions. The calculated curves shown in Figures 6.27 (b) and (c) were obtained by putting q = 10 and increasing the value of U_1 relative to

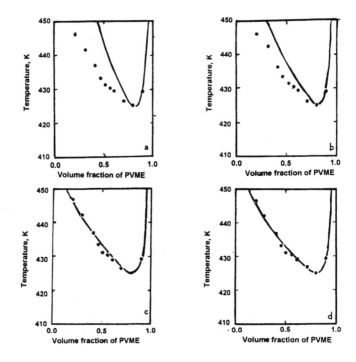

Figure 6.27. Comparison of experimental spinodal data and calculated spinodals for PVME/PS mixture. (Adapted, by permission from I.C. Sanchez and A.C. Balasz, *Macromolecules*, **22**, 2325 (1989)).

ε_{12} whilst decreasing the absolute value of ε_{12}. In (b), the ratio U_1/ε_{12} is 0.3, in (c) it is 0.5. It was found that the exact value of q is not important in establishing the ability of the model to predict the spinodal. In Figure 6.27 (d) another "best-fit" is shown at different parameter settings.

In Figure 6.28 experimental spinodal data of the much higher molecular weight H series sample are shown. In Figure 6.28 (a) the calculated spinodal was determined by using the same energy parameters as in Figure 6.27 (a), hence without specific interactions. The predicted effect of increase of molecular weight is clearly too large. An excellent prediction was obtained when the parameters used in Figures 6.27 (c) or (d) were used, hence including the effect of specific interactions. Specific interactions tend to stabilize the mixture and hence to reduce the effect of decrease of the LCST upon increase of molecular weight. The typical parameter setting of $-U_1/k$ around 300K

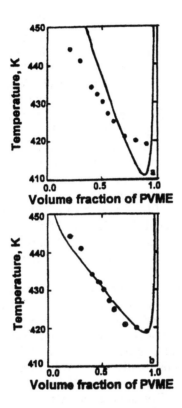

Figure 6.28. Comparison of experimental spinodal data of the H-series sample with predictions of the generalized LF model. (a) parameters as in Figure 6.27 (a). (b) parameters as in Figure 6.27 (c) or (d). (Adapted, by permission, form I.C. Sanchez and A.C. Balasz, *Macromolecules*, **22**, 2325 (1989)).

is equivalent to a specific interaction strength $U_1 = -0.6$ kcal/mol, still a relatively modest value when compared to the -5 kcal/mol typical for hydrogen bonds in low molecular weight liquids.[55]

Summarizing, analysis of the various models shows that the LCST demixing of PS/PVME mixtures can best be described and predicted using a model in which both compressibility and specific interactions are taken into account. It is difficult to judge in how far the one contribution is more important than the other. Irrespective of the model tested, the strength of the specific interaction was found to relatively weak, making the effect of compressibility more tangible.

6.3.2.2 Some typical polymeric mixtures

An example of a completely miscible blend with actual commercial importance is PS in polyphenyleneoxide (PPO).[56] Both materials are fully amorphous and the blends have single glass transition temperatures at all compositions. PS/PPO blends have a better balance of mechanical and processing properties than the constituents and are also known under the trade name Noryl. Neither an LCST, nor an UCST was ever observed. The fact that the solubility parameters of PS (9.1) and PPO (8.9) are closely matched is most likely the determining factor that brings about miscibility.

Another example of miscibility thanks to solubility parameter matching is polyisoprene (PIP) and polybutadiene (PBD).[57] It is a fully non-polar mixture and the solubility parameter difference (PIP: 8.2; PBD: 8.1) is within experimental error. An UCST was (evidently) never observed but an LCST was observed at a relatively low temperature of 338K. No model descriptions of the phase behavior are reported, but a mismatch in thermal expansivities between the constituents due to the CH_3 side group in PIP is most likely the driving force behind LCST demixing.

A special group of miscible blends is mixtures in which specific C=O...H-C-Cl Lewis acid-base interactions[58] are formed. Polyvinylchloride (PVC) was found to be miscible with polymethylmethacrylate (PMMA)[59] and with the series polypropylacrylate (PPrA), polybutylacrylate (PBA), and polypentylacrylate (PPeA).[8] LCSTs were observed at 463K, 403K, 398K, and 378K respectively. CPCs of high molecular weight and symmetric PVC – polyacrylics mixtures, as determined by turbidity measurements on cast films, were shown earlier, in Figure 6.2. Phase separation occurred in the order PPrA, PBA, PPeA. Low molecular weight oligomeric analogues of the same polymers were prepared as well. Experimental heats of mixing of these mixtures are shown in Figure 6.29. Interestingly, the same order of miscibilities is observed.

Solubility parameter differences between PVC and the polyacrylics investigated is almost similar for all three blends, namely 0.5 √(cal/cc). Hence, the consistency of decrease in LCST together with decrease in heats of mixing with a lower concentration of interacting groups points in the direction of domination of the phase behavior by specific interactions. Using a procedure similar to the generalization of the LF model for specific interactions by Sanchez *et al.*, Walsh *et al.*[60,61] also modified the Flory EoS theory in order to simulate spinodal curves for mixtures with specific interactions. It was proposed to calculate an effective (negative) interaction parameter χ from heats of mixing data of low molecular weight analogues. The specific interaction entropy parameter Q should than be adjusted to fit the calculated

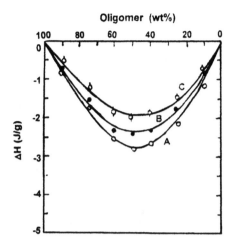

Figure 6.29. Experimental heats of mixing at 90.5°C of oligomeric analogue of PPrA (A), PBA (B), and PPeA (C) with oligo-PVC. (Adapted, by permission, from C.K. Sham and D. Walsh, *Polymer*, **28**, 957 (1987)).

spinodal to the minimum of the CPCs. It was found that such was possible for the PVC/polyacrylics mixtures and that the fit was relatively insensitive to the choice of pure component EoS parameters as well as to their differences. However, it was found that when the negative interaction parameters determined from heats of mixing were used the calculated spinodals were far too flat-bottomed and located even outside the binodal curve, see Figure 6.30. It was found that a much smaller negative value of χ (smaller by almost two orders of magnitude) generated a spinodal with much more curvature, thereby much better describing the experimental phase boundaries. The smaller value of χ needed also to be compensated for by a smaller value of Q. Hence, the phase behavior of the polymeric mixture could only be described satisfactory with a set of χ,Q parameters much smaller than those determined from oligomeric analogues. The relevance is that the specific interactions in the polymers tend to be less strong and less directional specific than in the chemically similar oligomers. Apparently, the polymeric building blocks involved in specific interactions tend to be more hindered in their degrees of freedom.

Another example of a miscible PVC blend is a mixture of PVC with polycaprolactone (PCL).[62] No cloud points were observed at all. The mixture is completely miscible over the whole temperature and composition range. The solubility parameter difference is not small (0.7). Apparently, specific interactions must be stronger than in the

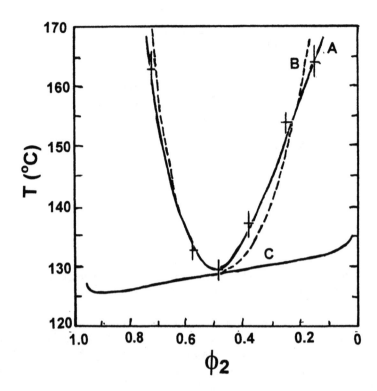

Figure 6.30. Simulated spinodal curves of PPaR/PVC blend. (A): experimental cloud point curve; (B) simulation using $\chi = -0.14\ \text{Jcm}^{-3}$ and $Q = -2.2 \times 10^{-4}$ $\text{Jcm}^{-3}\text{K}^{-1}$; (C) simulation using $\chi = -15.11\ \text{Jcm}^{-3}$ and $Q = -3 \times 10^{-2}\ \text{Jcm}^{-3}\text{K}^{-1}$. (Adapted, by permission, from C.K. Sham and D. Walsh, *Polymer*, **28**, 957 (1987)).

PVC/polyacrylics mixtures. A likely, but not proven reason for this is that in the polyacrylic side chains the -O-C=O- functional group is geometrically screened by the methyl group, whereas the carboxyl group in PCL is not.

A Lewis acid-base complex is formed in mixtures of PMMA and styrene-acrylonitrile (SAN) copolymer.[63] In this case, C=O..H-C-C≡N interaction is formed. The solubility parameter difference between PMMA, with 9.4 √(cal/cc) and SAN can be varied by means of the AN co-monomer content in the SAN copolymer. Complete miscibility over the full range of temperatures and compositions was observed at an AN-level of 13%. At this loading level solubility parameters are matched. The difference increases both when AN-loading is raised or lowered. In both cases, PMMA/SAN mixtures become less miscible.

LCSTs are observed and corresponding critical temperatures decrease with increasing solubility parameter difference. Complete immiscibility was observed at AN-levels lower than 9% and higher than 35%.

The above examples illustrate that in general the presence of specific interactions is a necessary, but not sufficient condition to bring about miscibility in high molecular weight polymeric mixtures. Simple dispersive interactions appear to be relatively important and in some cases the compressible nature of the mixture should be taken into account as well. Combination of these subtle effects makes quantitative description of the phase behavior extremely difficult. In order to raise the predictive power to a more quantitative level, it will above all be necessary to be able to predict strength and ordering effects of specific interactions in actual polymeric mixtures. With the aid of computer-added molecular modelling, classes of specific interacting groups and corresponding limiting solubility parameter differences may be identified.

REFERENCES

1. O. Olabisi, L.M. Robeson, and M.T. Shaw, *Polymer-polymer miscibility*, Academic, NY, 1979.
2. A. Einstein, *Ann. d. Phys.*, **33**, 1275 (1910).
3. M. von Smoluchowski, *Ann. d. Phys.*, **25**, 205 (1908).
4. L.D. Landau and E.M. Lifshitz, *Statistical Physics*, Pergamon, NY 1980.
5. H.E. Stanley, *Introduction to phase transitions and critical phenomena*, Oxford Un., NY 1971.
6. S. Saeki, N. Kuwahara, S. Konno, and M. Kaneko, *Macromolecules*, **6**, 246 (1973).
7. P.G. de Gennes, *Scaling concepts in polymer physics*, Cornell, London, 1979.
8. C.K. Sham and D. Walsh, *Polymer*, **28**, 957 (1987).
9. J.W. Cahn, *J. Chem. Phys.*, **42**, 93 (1964).
10. T. Hashimoto, J. Kumaki, and H. Kawai, *Macromolecules*, **16**, 641 (1983).
11. B. Berne and R. Pecora, *Dynamic light scattering*, J. Wiley, NY, 1976.
12. J.A. Stratton, *Electromagnetic theory*, McGraw-Hill, NY, 1941.
13. M.B. Huglin, *Light scattering from polymer solutions*, Academic, NY, 1972.
14. W. Burchard, *Adv. Polym. Sci.*, **48**, 1 (1983).
15. B.H. Zimm, *J. Chem. Phys.*, **16**, 1099 (1948).
16. G.D. Wignall in *Encyclopedia of Polym. Sci. and Eng.*, 2nd edition, Vol. 10, p.112, Wiley, NY, 1987.
17. I.I. Gurevitch and L.V. Tarasov, *Low Energy Neutron Physics*, North-Holland Publishing, Amsterdam, 1968.
18. R.J. Young, *Introduction to polymers*, Chapman and Hall, London, 1981.
19. R.N. Howard, ed., *The Physics of glassy polymers*, Appl. Sci. Publ., London, 1973.
20. A. Ledwith and A.M. North, ed., *Molecular behavior and the development of polymeric materials*, Wiley, NY, 1975.
21. A.W. Birley, B. Haworth, and J. Batchelor, *Physics of plastics*, Oxford Un., NY, 1991.
22 T.G. Fox, *Bull. Am. Phys. Soc.*, **68**, 441.
23 I. Hashizume, A. Teramoto, and H. Fujita, *J. Polym. Sci., Polym. Phys. Ed.*, **19**, 1405 (1981).

24. Th.G. Scholte, *J. Polym. Sci., Part A*, **9**, 1553 (1971).
25. K.W. Derham, J. Goldsbrough, and M. Gordon, *Pure & Appl. Chem.*, **38**, 97 (1974).
26. I.H. Park, Q.W. Wang, and B. Chu, *Macromolecules*, **20**, 1965 (1987).
27. J.M.G. Cowie, A. Maconnachie, and R.J. Ranson, *Macromolecules*, **4**, 57 (1971).
28. S. Saeki, N. Kuwahara, M. Nakata, and M. Kaneko, *Polymer*, **16**, 445 (1975).
29. B.A. Wolf and H. Gaerisen, *Coll. & Polym. Sci.*, **259**, 1214 (1981).
30. K.W. Derham, J. Goldsbrough, and M. Gordon, *Pure Appl. Chem.*, **38**, 97 (1974).
31. P. Irvine and M. Gordon, *Macromolecules*, **13**, 761 (1980).
32. R. Koningsveld, L.A. Kleintjes, and A.R. Schultz, *J. Polym. Sci., Part A2*, **8**, 1261 (1970).
33. E. Nies and A. Stroeks, *Macromolecules*, **23**, 4088 (1990).
34. A. Stroeks and E. Nies, *Macromolecules*, **23**, 4092 (1990).
35. M.L. Huggins, *Ann. N.Y. Acad. Sci.*, **44**, 431 (1943).
36. A. Quach and R. Simha, *J. Appl. Phys.*, **42**, 4592 (1971).
37. J. Jonas, D. Hasha, and S.G. Huang, *J. Phys. Chem.*, **84**, 109 (1980).
38. K.S. Siow, G. Delmas, and D. Patterson, *Macromolecules*, **5**, 29 (1972).
39. I.C. Sanchez and R.H. Lacombe, *Macromolecules*, **11**, 1145 (1978).
40. J.M. Bardin and D. Patterson, *Polymer*, **10**, 247 (1969).
41. J.B. Wedgeworth and C.J. Glover, *Macromolecules*, **20**, 2268 (1987).
42. T. Nose and B. Chu, *Macromolecules*, **12**, 1122 (1979).
43. R. Kirste and G. Schultz, *Z. Phys. Chem. NF*, **27**, 20 (1960).
44. M.A. van Dijk and A. Wakker, *Polymer*, **34**, 132 (1993).
45. A. Wakker, F. van Dijk, and M.A. van Dijk, *Macromolecules*, **26**, 5088 (1993).
46. C.C. Han, M. Okada, Y. Muroga, B.J. Bauer, and Q. Tran-Cong, *Polym. Eng. Sci.*, **26**, 1208 (1986).
47. M.G. Brereton, E.W. Fisher, C. Herkt-Maetzky, and K. Mortensen, *J. Chem. Phys.*, **87**, 6144 (1987).
48. C.C. Han, B.J. Bauer, J.C. Clark, Y. Muroga, Y. Matsushita, M. Okada, Q. Tran-Cong, T. Chang, and I. Sanchez, *Polymer*, **29**, 2002 (1988).
49. L.P. McMaster, *Macromolecules*, **6**, 760 (1973).
50. J. Kumaki and T. Hashimoto, *Macromolecules*, **19**, 763 (1986).
51. A detailed explanation of the random phase approximation can be found in Reference 7.
52. G. ten Brinke and F.E. Karasz, *Macromolecules*, **17**, 815 (1984).
53. A. Wakker and M.A. van Dijk, *Polym. Networks Blends*, **2**, 123 (1992).
54. I.C. Sanchez and A.C. Balasz, *Macromolecules*, **22**, 2325 (1989).
55. R.E. Goldstein and J.S. Walker, *J. Chem. Phys.*, **78**, 1492 (1983).
56. See e.g. Reference 1 Section 5.2.2.
57. C.A. Trask and C.M. Roland, *Polym. Comm.*, **29**, 332 (1988).
58. W.B. Jensen, *The Lewis acid-base concepts*, Wiley, NY, 1980.
59. E.J. Vorenkamp and G. Challa, *Polymer*, **28**, 957 (1987).
60. D.J. Walsh and S. Rostami, *Makromol. Chem.*, **186**, 145 (1985).
61. D.J. Walsh and S. Rostami, *Adv. Polym. Sci.*, **70**, 119 (1985).
62. E. Nolley, D.R. Paul, and J.W. Barlow, *J. Appl. Polym. Sci.*, **23**, 623 (1979).
63. M. Suess, J. Kressler, and H.W. Kammer, *Polymer*, **28**, 957 (1987).

List of Symbols

Symbol	Meaning	Introduced in:

Chapter 2

Symbol	Meaning	Introduced in:
β	bond index	2.5.2
β^v	valence bond index	2.5.2
c_i	concentration of species i	2.1.3
C	number of components (species)	2.1.3
δ	solubility parameter	2.5
δ	connectivity index	2.5.2
ΔG_M	Gibbs free energy of mixing	2.2.3
ΔH_M	enthalpy of mixing	2.2.3
$\Delta \mu_i$	chemical potential difference between mixture and pure phase of species i	2.2.3
ΔS_M	entropy of mixing	2.2.3
ΔV_M	volume change on mixing	2.2.3
γ	surface tension	2.1.2
F	Helmholtz free energy	2.1.2
Φ_i	volume fraction of species i	2.1.3
Φ	composition (described in volume fractions)	2.2.4
Φ^F	overall composition (divided over phases)	2.2.4
Φ_c	critical composition	2.2.5
G	Gibbs free energy	2.1.2
G^V	Gibbs free energy per unit volume	2.1.3
χ	interaction parameter	2.4
$^0\chi$	zeroth order atomic connectivity index	2.5.2
$^0\chi^v$	zeroth order valence atomic connectivity index	2.5.2
$^1\chi$	first order atomic connectivity index	2.5.2
$^1\chi^v$	first order valence atomic connectivity index	2.5.2
H	enthalpy	2.1.2
H_f	heat of fusion (crystallization)	2.2.7
P	pressure	2.1.2
k	Boltzmann's constant	2.3
λ_i	activity of species i	2.1.3
M_i	molar mass of species i	2.1.3
μ_i	chemical potential of species i	2.1.3
n_i	number of molecules of species i	2.1.3
ν	phase volume fraction	2.2.4
Ph	number of phases in equilibrium	2.2.2
Π	osmotic pressure	2.3
R	molar gas constant	2.1.3
ρ_i	mass density of species i	2.1.3
S	entropy	2.1.1
S_f	entropy of fusion (crystallization)	2.2.7
T	temperature	2.1.1
T_c	critical temperature	2.2.5
U	energy	2.1.2

199

Symbol	Meaning	Introduced in:
V	volume	2.1.2
V_i	partial molar volume of species i	2.1.3
V_m	molar volume	2.5
W_i	mass fraction of species i	2.1.3
x	composition (described in mole fractions)	2.2.4
x_i	mole fraction of species i	2.1.3
I	superscript denotes phase I	2.2.1
II	superscript denotes phase II	2.2.1

Chapter 3

A_2	second virial coefficient	3.5
b_k	Kuhn statistical length	3.2.1
C_∞	characteristic ratio	3.2.1
Γ	interaction function	3.6.1
D	dimension	3.2.2
F	superscript denotes Feed (overall composition)	3.6.1
Φ	volume fraction	3.6.1
χ	interaction parameter	3.6.1
χ_h	enthalpic contribution to χ	3.6.1
M	molar mass	3.1.1
M_n	number averaged molar mass	3.1.1
M_w	weight (mass) averaged molar mass	3.1.2
M_z	z-averaged molar mass	3.1.2
r	number of segments per chain	3.3.2
r_w	weight averaged number of segments per chain	3.3.2
r_z	z averaged number of segments per chain	3.3.2
R_{ee}	end-to-end distance	3.2.1
$<R_{ee}^2>$	root mean square average end-to-end distance	3.2.1
$<s^2>$	radius of gyration	3.2.1
W_i	mass fraction of component I	3.1.2
ψ_i	volume fraction of polymers with chain length r_i within one 'family'	3.6.1
Ω_{conf}	number of configurations	3.2.1

Chapter 4

$A(N_1..N_k)$	number of ways of k loops with N_k bonds	4.5
α_i	thermal expansivity component i	4.3.2
β	temperature parameter to χ	4.2
β	$\beta = (kT)^{-1}$	4.3.1
c	chain flexibility parameter	4.3.1
$c(r)$	direct pair correlation function	4.5
$C(r)$	site matrix of direct pair correlation function	4.5
χ	interaction parameter	4.2
χ_c	critical value of χ	4.2
χ_h	enthalpic part to χ	4.2
χ_s	entropic part to χ	4.2
χ_{disp}	dispersive part to χ	4.3.2
χ_{spec}	specific interaction part to χ	4.4
χ_{free}	free volume part to χ	4.3.2

Symbol	Meaning	Introduced in:
δ_i	solubility parameter component i	4.2
$\Delta\delta$	solubility parameter difference	4.2
$E_i(0)$	segmental lattice energy at equilibrium	4.3.2
$\varepsilon(r)$	generalized potential energy function	4.3.1
ε^*	energy scale factor to ε	4.3.1
Φ_i	volume fraction component i	4.2
F	Helmholtz free energy	4.3
F_{conf}	configurational free energy	4.3.1
$g(r)$	atomic radial distribution function	4.5
$g_{\alpha\beta}(r)$	site-site pair distribution function	4.5
ΔG_M	Gibbs free energy of mixing	4.2
γ	non-ideality factor to χ	4.2
γ	geometrical factor	4.3.1
h	Planck's constant	4.3.1
h	external interaction field	4.5
k	Boltzmann's constant	4.2
k	wavevector	4.5
κ_T	isothermal compressibility	4.5
λ	fraction of specific interactions	4.4.1
N	total number of system molecules	4.3.1
N_A	Avogadro's number	4.2
n_e	number of external degrees of freedom	4.3.1
n_i	number of internal degrees of freedom	4.3.1
p	pressure	4.3
$\underline{p}$	reduced pressure	4.3.1
p^*, P^*	pressure reduction parameter	4.3.2
$\mathbf{p}_i$	impulse of molecule i	4.3.1
Q	configurational partition function	4.3.1
Q	adjustable volume parameter of cell partition function	4.3.2
q	number of specific interaction sites	4.4.1
R	gas constant	4.2
ΔS_M	entropy of mixing	4.3.3
ΔS_{comb}	combinatorial entropy of mixing	4.3.3
ΔS_{holes}	holes entropy of mixing	4.3.3
r^*	position scale factor to ε	4.3.1
r_i	chain length component i	4.2
$\mathbf{r}_i$	position of molecule i	4.3.1
$\bar{\rho}$	reduced density of mixture	4.3.3
S_i	magnetic spin of lattice site i	4.5
$S(k)$	structure factor	4.5
σ	molecular surface area	4.2
T	temperature	4.2
$\bar{T}$	reduced temperature	4.3.1
T^*	temperature reduction parameter	4.3.2
τ	characteristic free volume difference	4.3.2
U	potential energy	4.3.1
U_1	specific interaction energy	4.4.1
U_2	dispersive interaction energy	4.4.1
V	volume	4.3
V_L	lattice site volume	4.2

Symbol	Meaning	Introduced in:
V_{lat}	molar lattice volume	4.4.1
v	reduced volume	4.3.1
v^*, V^*	volume reduction parameter	4.3.2
v_0	core volume	4.3.1
$\omega(r)$	mean energy of interaction	4.3.1
Ω	volume of spins phase space	4.5
$\Omega(r)$	site matrix of intramolecular distribution function	4.5
Ψ	cell partition function	4.3.1
z_i	lattice coordination number component i	4.2
Z	partition function	4.3.1
Z_{tr}	translational degrees of freedom to Z	4.3.1
Z_i	internal degrees of freedom to Z	4.3.1

Chapter 5

E	energy	5.2.1
ϕ	volume fraction	5.6.1
G	Gibbs free energy	5.2.3
h	Planck's constant	5.2.3
m	mass of a particle	5.2.3
μ	chemical potential	5.2.1
N	number of particles	5.2.1
P	pressure	5.2.1
P	P(X), property	5.2.1
q	partition function	5.2.2
Q_{NVE}	partition function of microcanonical ensemble	5.2.1
Q_{NVT}	partition function of canonical ensemble	5.2.1
Q_{NPT}	partition function of isothermal-isobaric ensemble	5.2.1
$Q_{\mu VT}$	partition function of grand canonical ensemble	5.2.1
R_{ee}	end-to-end distance	5.6.1
X	phase-space coordinate	5.2.1
V	volume	5.2.1
W	number of ways to realize a distribution	5.2.2
Z	configurational partition function	5.3

Chapter 6

A_2	osmotic second virial coefficient	6.2.1
b	scattering length	6.2.1
c	concentration	6.2.1
$<dc^2>$	concentration fluctuations, mean square	6.2.1
$dc(q,t)$	concentration fluctuations; wavevector, time dependent	6.2.1
c	chain flexibility parameter	6.3.1
χ	interaction parameter	6.3.1
χ_c	critical value of χ	6.3.1
χ_h	enthalpic part to χ	6.3.1
χ_s	entropic part to χ	6.3.1
δ	solubility parameter	6.3.1

Symbol	Meaning	Introduced in:
ε^*	energy scale parameter HH, S&S models	6.3.1
$<d\varepsilon(q)^2>$	mean square dielectric constant fluctuations, wavevector dependent	6.2.1
ϕ_i	volume fraction component i	6.3.1
ΔG_M	Gibbs free energy of mixing	6.2.1
G^V	free energy per unit volume	6.2.1
$I(q,t)$	wavevector and time dependent light scattering intensity	6.2.1
k_B	Boltzmann's constant	6.2.1
K_*	constant	6.2.1
K^*	constant	6.2.1
k	constant	6.2.1
λ	wavelength	6.2.1
M_{AB}	mobility between polymers A,B	6.2.1
M_w	molecular weight, weight average	6.2.1
M_n	molecular weight, number average	6.3.1
n	refractive index	6.2.1
N_A	Avogadro's number	6.2.1
$P(q)$	form factor, wavevector dependent	6.2.1
q	wavevector	6.2.1
q_c	critical value of q	6.2.1
q_{max}	value at which $R_c(q)$ is maximal	6.2.1
q	scattering angle	6.2.1
r	chain length	6.3.1
$R(q)$	Rayleigh ratio, wavevector dependent	6.2.1
R_o	isotropic Rayleigh ratio	6.2.1
R_m	measured Rayleigh ratio	6.2.1
$R_c(q)$	relaxation rate of concentration fluctuations	6.2.1
$<R_g^2>_z$	radius of gyration, mean square, z-average	6.2.1
$<d\rho^2>$	density fluctuations, mean square	6.2.1
$S(q)$	structure factor, wavevector dependent	6.2.1
S_D	single chain form factor	6.3.1
σ	scattering cross-section	6.2.1
t	time	6.2.1
T	temperature	6.2.1
T_g	glass transition temperature	6.2.4
τ	turbidity	6.2.1
V	scattering volume	6.2.1
V^*	volume scale parameter HH, S&S models	6.3.1
v_i	molar volume component i	6.3.1
ξ	correlation length	6.3.1
z	lattice coordination number	6.2.1

Index